I0815479

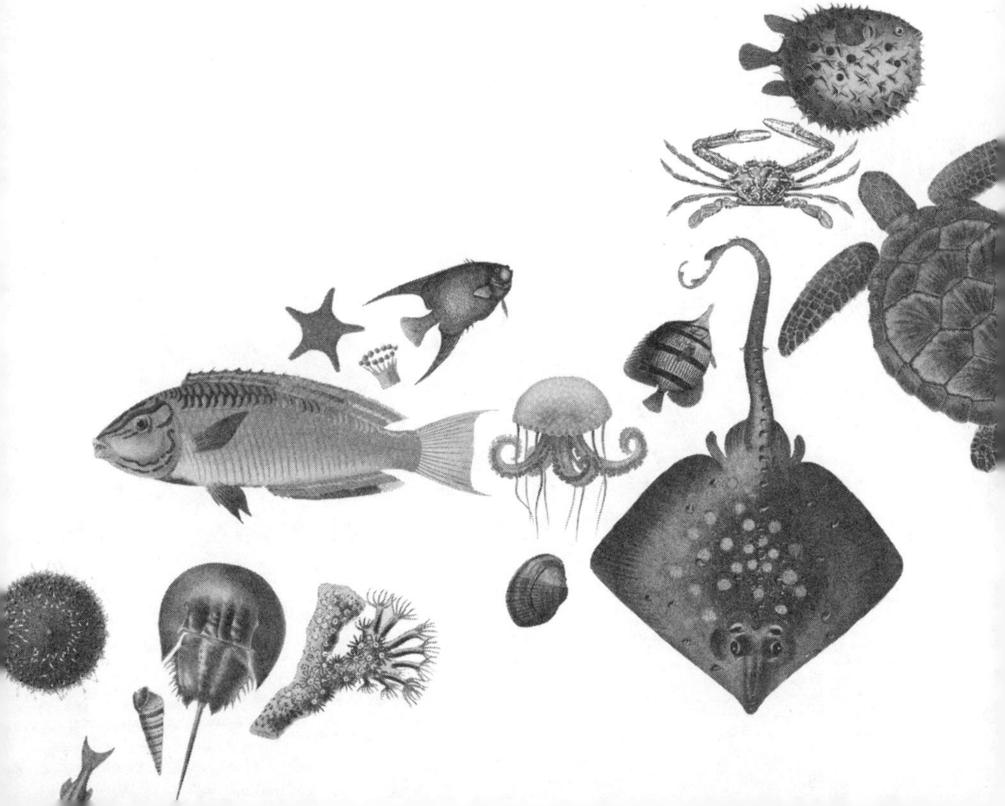

Rachel Jordan is a marine biologist and spiritual guide. Using the ocean as a lens, she points us to the God revealed in biology. Her writing is as rich and colorful as the sea creatures she describes. Her love for the ocean is a clear reflection of God's.

John Van Sloten, theologian and author of *God Speaks Science: What Neurons, Giant Squid, and Supernovae Reveal About Our Creator*

Rachel Jordan has done us a tremendous service with her presentation of theology and ecology in light of our aching search for truth. Rachel is uniquely gifted in both areas of study and assures us that there are indeed not as many conflicts between science and Scripture as is commonly assumed. Readers will be fascinated by the enchanting descriptions of creation and motivated to keep seeking truth—and I trust they'll find assurance in the conclusions reached in *If the Ocean Has a Soul*.

Peter Reid, former principal of Torchbearer Bible School Bodenseehof and general director of Torchbearers International

Marine biologist and devoted Christian Rachel G. Jordan has crafted an exceptional book that seamlessly integrates her vibrant scientific expertise and her faith in a way that's both enlightening and deeply personal. Without forcing conclusions, she presents a wealth of resources and insights, inviting readers to think critically and ask hard questions. *If the Ocean Has a Soul* explores Scripture and connects it to creation—from vast coral reefs to leopard sharks to creatures as small as hermit crabs and as disregarded as betta fish. You'll feel like you're right there with her, learning side by side. Ocean lovers, this is a must-read. Christians, come learn. And to those exploring their faith, don't pass up Rachel's thoughtful insights.

Marissa Myer, executive director of The Coral Nursery of Puerto Rico

As a pastor and teacher for over four decades, I have met countless people who felt that science and faith are incompatible. For anyone dealing with such matters, Rachel Jordan's book is a gift. She weaves science, faith, and her own story together as a seamless ecosystem, inviting people to curiosity, awe, worship, and even wholeness in the process. A must-read!

Richard Dahlstrom, author, teacher, and conference speaker

This book is a deeply personal and compelling account of the author's journey as a Christian and marine biologist. She poignantly expresses her awe at God's creation, her grief over the loss of her beloved coral, and her hope for creation's resilience. The resounding message of this book is that we need both faith and scientific knowledge to fully understand and to be effective stewards of God's creation.

Jennifer Tenlen, PhD, associate professor of biology at Seattle Pacific University

if the ocean has a soul

TYNDALE
REFRESH®

Think Well. Live Well. Be Well.

RACHEL G. JORDAN

if the ocean has a soul

a marine biologist's pursuit of truth through deep waters of faith and science

Visit Tyndale online at tyndale.com.

Visit Tyndale Refresh online at tyndalerefresh.com.

Visit Rachel at rachelgjordan.com.

Tyndale, Tyndale's quill logo, *Tyndale Refresh*, and the Tyndale Refresh logo are registered trademarks of Tyndale House Ministries. Tyndale Refresh is a nonfiction imprint of Tyndale House Publishers, Carol Stream, Illinois.

If the Ocean Has a Soul: A Marine Biologist's Pursuit of Truth through Deep Waters of Faith and Science

Copyright © 2025 by Rachel G. Jordan. All rights reserved.

Cover illustration copyright © Jago. All rights reserved.

Interior illustration of coral icon copyright © BUSAIRI/TheNounProject.com. All rights reserved.

Interior illustration of shell icon copyright © mas r/TheNounProject.com. All rights reserved.

Author photo by Brenda Johns, copyright © 2021. All rights reserved.

Cover design by Julie Chen

Interior design by Laura Cruise

Published in association with the literary agency of Gardner Literary LLC, gardner-literary.com.

Unless otherwise indicated, Scripture quotations are taken from the *Holy Bible*, New Living Translation, copyright © 1996, 2004, 2015 by Tyndale House Foundation. Used by permission of Tyndale House Publishers, Carol Stream, Illinois 60188. All rights reserved.

Scripture quotations marked BSB are taken from The Holy Bible, Berean Study Bible, BSB. Copyright © 2016, 2018 by Bible Hub. Used by permission. All rights reserved worldwide.

Scripture quotations marked ESV are from The ESV® Bible (The Holy Bible, English Standard Version®), copyright © 2001 by Crossway, a publishing ministry of Good News Publishers. Used by permission. All rights reserved.

Scripture quotations marked NIV are taken from the Holy Bible, *New International Version*,® *NIV*.® Copyright © 1973, 1978, 1984, 2011 by Biblica, Inc.® Used by permission. All rights reserved worldwide.

Scripture quotations marked RSV are taken from the Revised Standard Version of the Bible, copyright © 1946, 1952, and 1971 National Council of the Churches of Christ in the United States of America. Used by permission. All rights reserved worldwide.

The URLs in this book were verified prior to publication. The publisher is not responsible for content in the links, links that have expired, or websites that have changed ownership after that time.

For information about special discounts for bulk purchases, please contact Tyndale House Publishers at csresponse@tyndale.com, or call 1-855-277-9400.

Library of Congress Cataloging-in-Publication Data

A catalog record for this book is available from the Library of Congress.

ISBN 979-8-4005-0584-3

Printed in the United States of America

31 30 29 28 27 26 25
7 6 5 4 3 2 1

For the ocean lovers, question askers,

and Truth seekers

contents

PREFACE

unfathomable seas

My eyes focus on the still surface of the water, reflecting gray coastal sky. This tide pool is just deep enough that my ankles would get wet if I stood in it. But I would never step into one because they are filled with beauty and life. Tide pools serve as temporary homes to creatures that get caught in crevices of rock when the tide goes out, washing over long rifting patches of land that trap them in puddles as the waters recede. Like pockets of the vast ocean, each pool holds its little wonders.

As I carefully dip my hand into the water, ripples fan out over its small surface, distorting reflections. And suddenly, there are the creatures. Glossy red threads of marine algae. Pebble-like shells of striped periwinkle snails. Clusters of clamped mussels so darkly purple they seem black. A brisk-walking fiddler crab, with one arm bigger than the other. Fuzzy green, Jell-O-looking sea anemones

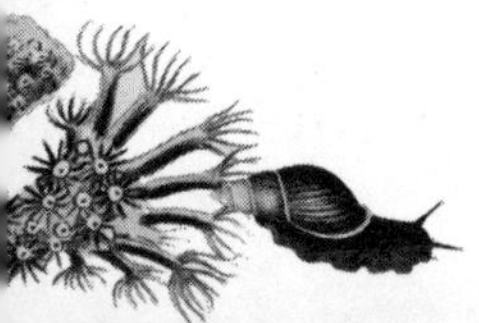

with sticky tentacles. The occasional jellyfish, translucent like the moon on a clear night. Even a rough-textured orange sea star, clutching tightly to the sides of a pool with its tender but amazingly strong tube feet.

Tide pooling is a bit like searching for truth. Both take intentionality, gentleness, and time. Nets and buckets are helpful for tide pooling, and maybe there are tools that can help us find truth too—tools like good books and trustworthy people. Just as living creatures can be hurt by rough handling, there is a risk of harm if we do not handle new concepts, ideas, and thoughts with care. Sometimes critical thinking can feel like sweaty labor while squatting on sharp rocks under the sun, but if we are willing to do it, we will find it well worthwhile. As we pull up unique, surprising, and biodiverse things, we might even come away feeling as though we have discovered treasures.

Both faith and science are necessary for truth-seeking because exposure changes perspective. What can be seen from the shoreline is different from what can be seen from the edge of a boat, which is still different from what can be seen sixty feet below the water's surface. If we live our whole lives on the shore, what might we miss?

Faith moves science off-land. It takes all the observational and experimental processes, scoops them up, and runs them out to sea. Faced against what are sometimes giant waves and what at other times is flat calm, faith forces science to grapple with that which cannot be observed with the naked eye or experimented on with even the greatest of theoretical cunning. Science is limited to what can be quantified and qualified, felt with assured evidence and firm as the ground underfoot. But like a boat speeding out to open sea, faith pushes beyond these limitations. Boundless and free, it brushes the wind through our hair and sprays salt on our skin. It explains what science cannot, touching the soul and guiding us along life's journey. But just as science without faith is landlocked, faith without science floats on the surface.

My own interest in marine science was piqued at a young age. Growing up making annual family pilgrimages to the Washington and Oregon coasts, I learned the ocean first through its tide pools. These pockets of briny water held colorful treasures that I longed to explore with my eyes and poke with my fingers. Brushing back tufts of algae from the pools' edges, I quickly found that crabs pack a pinch, sea anemones stick and squirt, and sea urchins poke and pierce. But despite all the tingling soreness in my curious fingertips, I was entranced by the complex and colorful creatures I discovered there.

The lessons I learned from tide pools took longer to translate to my life as a young Christian, however. Chief among these lessons was the realization that science takes faith deeper. With all its questions, doubts, and critical thinking, science forces penetration into the depths. Science does not allow faith to stagnate with surfacy ideals, trite religious comments, and immature beliefs. Like a weight belt on a scuba diver, science pulls faith into amazing, otherworldly, and challenging places where we might otherwise never choose to go, places that provide incredible opportunities to learn and grow. I felt this tug as a girl kneeling beside a tide pool, wondering what other creatures lurked in the ocean before me. You might say my training as a marine biologist—which has taken me to the reefs of Australia and southernmost Florida—began right there. Even then, I was beginning to understand that science engages us in conversation with the world, inviting us to commune with people and creatures as diverse as the fish in the sea.

For me, that conversation expanded exponentially while I served as the lead coral biologist at a national park situated on an island seventy miles from Key West—closer to Cuba than to the United States mainland. My work at the park allowed me to learn by application that which faith teaches in theory. But it also reinforced my belief that science can no more do the job of faith than faith can do that of science; we need both.

I should clarify that when I refer to faith throughout this book, I'm talking specifically about the Judeo-Christian tradition. The reasons for this are fourfold. First, I write from the perspective of a practicing Christian because I am one. Any self-aware Christian contemplating science is bound to encounter questions about whether faith and science can (or even should) relate. As a professional marine biologist who loves God, I find myself uniquely positioned to encourage such people. For the early career scientist who is also contemplating Christianity, I am here to address the overlapping dialogue, that internal push and pull between being a scientist in training and a disciple in training.

Second, Christianity (which began as an outgrowth of Judaism) teaches that the physical is as sacred as the spiritual. The creation of the world and humankind was not some heavenly accident or cosmic sneeze from an aloof divine being. Christianity and Judaism are unique among religions in that they declare that humankind was created for relationship with God.

What's more, while the early Christian church made significant contributions to the development of the biological sciences, there have been periods throughout history of what might be called persecution of the sciences. It is both common knowledge and unfortunate fact that modern-day Christianity is frequently poised in opposition to the advances of modern science. This must be addressed.

Finally, Christian faith and science seek the same thing—to know Truth. I've capitalized the *T* on *Truth* here because both faith and science pursue what is ultimately true—the biggest, realest, most actual Truth. From a faith understanding, this Truth points directly to God. From a scientific understanding, this Truth is always the goal, the objective toward which every theory and principle is aimed, the end to which every hypothesis and experiment is directed.

Because of this, the evidence, examples, and explanations provided through the stories in this book come from a diverse set of sources, with credible research and helpful quotations from Christians and atheists alike. Beyond seeking to identify what is true, I have not discriminated among sources. Every inclusion is trustworthy within its realm of study and communicates a useful piece of relevant wisdom. Scripture, along with Bible-believing theologians, pastors, and trustworthy leaders in the faith, is regularly quoted. Similarly, long-standing scientific theories, principles, and examples are referenced, along with direct quotes from scientists who are leaders in their fields.

With such a wealth of sources to consult, let me briefly clarify some terminology. When referring to the physical world, scientists tend to use the word *nature*, while Christians often use the word *creation*. As both a scientist and a Christian, I use these words interchangeably, but it is important to note that "in the Judaeo-Christian tradition, the word 'creation' has a broader meaning than 'nature,' for it has to do with God's loving plan in which every creature has its own value and significance. Nature is usually seen as a system that can be studied, understood and controlled, whereas creation can only be understood as a gift from the outstretched hand of the Father of all, and as a reality illuminated by the love which calls us together into universal communion."[1]

For our purpose of understanding the integration of faith and science from a marine biology–based perspective, we will lean into these definitions. I reserve the term *nature* for talking about the created world from the perspective of raw science, void of faith-based implications. I use *creation* to describe the natural, created world, which includes humans. Furthermore, I refer to *humankind* to describe humanity as a whole, including humans across space and time.

Science and faith may do different things along their search for Truth, but like the water that flows between low and high tide, we need them both. To judge one without the other is to forsake half the wonders we are meant to experience. While this collection of stories is centered around the unique experiences I have had in my professional scientific career, it is grounded in what the Creator—God—has taught me through those experiences about the world he created, the meaning behind it, humankind's purpose and place, and where everything is heading. This is a deep dive into ecology and theology, meant to explore the spaces where sacred and supposed secular intertwine. My intention, both in this piece of writing and in life, is to explore the integration of faith and science from the perspective of a Jesus-loving coral biologist with the goal of knowing the Creator and his creation better. While these stories may serve as simple inspiration or entertainment to anyone who loves nature and is curious about the reality of life as a marine biologist, the discussion of these experiences courses deeper, being ultimately an exegesis of marine biology.

We are like creatures living within tide pools, trapped in the intertidal zone between the water's edge and high tide line. Perhaps we have lost sight of the ocean. Perhaps there is something greater than the tiny pool where we currently reside. We long to know that God, the one who made and sustains our world, is active in each of our personal lives and has a good plan. As the tide sweeps in and out in frothing sea-foam and steady rhythmic time, he encourages us to explore. We can ask questions, conduct research, and consider thoughts and doubts that perhaps we never have before. It may be scary. It may even be a long journey. To a creature who has spent its whole life in a tide pool, the ocean would indeed seem scary and far away. And yet we wait at the shoreline, ready to be intentional, gentle, and brave, and to take our time.

As we figuratively run out to sea and dive deeper, we will rejoice, weep, and labor alongside the created world. We will wrestle with hard questions of meaning, life, suffering, and death. We will discover that, even when we can't currently see it, God is actively working to redeem and restore not only his creation but each of our stories. We will consider both biblical wisdom and scientific discovery as means of understanding the unique life experiences provided to a marine biologist. And ultimately, as we seek to understand these unfathomable seas, we will discover the character of God, who "made heaven and earth, the sea, and everything in them" (Psalm 146:6).

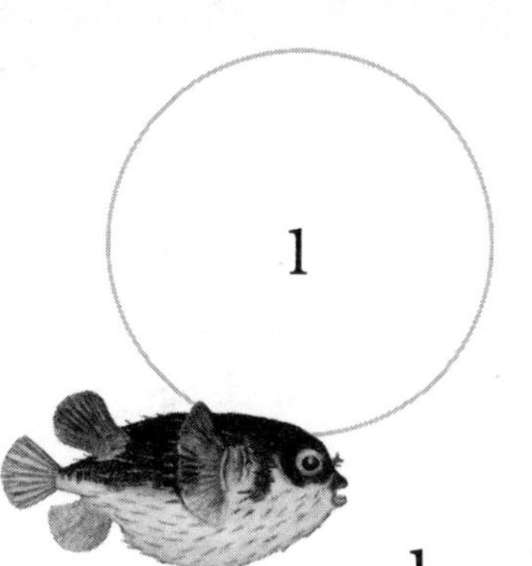

1

land of the living

One of my longtime friends lived at the Windjammer. When he saw you, he'd meander out from his little dwelling in the cleft between two large brain corals. You'd notice his pudgy eyes first, inky black with iridescent purple irises. His mouth protruded in a miniature curve, always a gentle smile. His patterned tan-brown body bulged in the front, tapering toward the back. From the sides, two little fins whirligiged through water. He puttered along, a perfect porcupinefish. I called him Porky.

We were both surprised when we first saw each other—he so doggedly sleepy from the cozy nap he'd been taking, me from the distraction of the enormous corals that created his home. It was my first trip to the Windjammer, and I was awestruck by the tremendous colonies growing on the vertical hull of the shipwreck and the torrents of brilliantly flashing amberjacks schooling before me. A large

pink crab stealthily scuttled over the encrusted debris, clipping his claws at a pair of territorial purple damsels before vanishing into a hole. Pointy-nosed needlefish danced jaggedly under the surface where the water crested over the tip of the wreck, breaking down in bubbly explosions. A goliath grouper nearly the size of our dive boat haunted the moving shadows below, his faint outline a contrast to the flashes of patrolling barracuda, silver and toothy. Amazingly textured corals colored emerald, ruby, amber, amethyst, and sapphire gleamed in the shallow sunlight. Over yonder, purple sea fans beat in slow motion with the current, invisible vibrations shivering their intricate branches, like leaves of a tree lightly rustling in wind. As fish darted and gathered and rested, fronds of the soft coral thicket swayed left then right, beating back and forth to a never-ceasing rhythm of life. It was all a marvelous wonder to me. For Porky, it was just another day in the neighborhood.

The Windjammer is unique not only for the incredible plethora of creatures inhabiting the site but because of the shipwreck itself. Having sunk in 1907 while en route from Florida to Uruguay, the 261-foot *Avanti* was discovered in Dry Tortugas National Park in 1971 by an archaeological survey crew.[1] It took several decades to identify the wreck, during which time the site was named the Windjammer. While the cause of the wreck is unknown, the bow and stern were severed and lie adjacent to each other over several hundred feet of sandy habitat, with an array of midship wreckage jumbled between them, which creates plenty of interesting focal points for snorkelers and divers.

Heavily encrusted with corals, the Windjammer serves as the substrate for a remarkable array of creatures. After surfacing from our first dive at the site, my diving buddy proclaimed, "That is the most coral-y dive I've ever done. Those were the biggest coral colonies I have ever seen!" The enthusiasm was heartfelt. The vibrant colors and ever-changing movements of the wreck's community were beyond-words beautiful. The corals made our jaws fall open. The fish made us want to laugh. The structure of the wreck was fascinating. Something about the Windjammer moved a diver's soul.

The site was beloved. We went back time and again, always seeing something new, yet always seeing Porky—happily living as the solitary porcupinefish at the wreck. I loved to check up on that little fish. He'd scoot out of his cleft between the corals, waggle his tiny yellow fins at me, and then leisurely cruise over to the sea fan field. But I wasn't just diving at the Windjammer for fun; it was my job.

I was performing disease reconnaissance—swimming with the flow of the current, checking every coral within five feet on either side of my path, and documenting data on the makeshift clipboard I held. My eyes scanned each coral for the stark white lesions indicative of a notoriously violent disease, stony coral tissue loss disease (or SCTLD, for short), which had been sweeping along the Florida Reef and Caribbean, rapidly killing twenty-one coral species in its wake.[2]

Any reliable ecologist will tell you that disease is an expected and integral part of healthy ecosystems. A disease develops when an organism (parasite or pathogen) manipulates the body and function of another organism (the host) within an environment. Similar to how periodic wildfires can be beneficial, adding nutrients back into soil and clearing out cluttered habitat space, a moderate amount of coral disease has the power to regenerate and revitalize a reef.

In many ways, the environment dictates the success or failure of a disease to originate, intensify, and spread. In an environment where everything is well-connected (like the ocean, where water is ever flowing and ever mixing), transmission rates can spike. When the environment is primed with the perfect set of conditions, and potential hosts stand vulnerably exposed, it is only a matter of time before parasites and pathogens take hold, leading to a more widespread, impactful, and potentially deadlier disease—a true disease "outbreak" or event. Out of balance and disproportionate to the function of their environments, disease outbreaks are ecologically harmful. In recent years, humankind became personally acquainted with the intensity and transmissibility of a coronavirus disease outbreak, as the virulent COVID-19 discovered its perfect niche and created a global pandemic. I once heard someone say that SCTLD is the COVID of corals. They were wrong; SCTLD is far more deadly, killing over half of infected corals.[3]

Considered an "unprecedented" coral disease outbreak for its expanding geographic cover, number of species impacted, and tremendously high mortality rates, scientists were in a frenzy to determine both SCTLD's cause and cure.[4] Because the disease originated near Miami in 2014, it has been hypothesized that a long-buried pathogen was unearthed during standard port dredging.[5] It could be argued that the pathogen causing SCTLD has always been present in the environment, mildly free-roaming, existing on minute and unnoticeable scales. Perhaps the environmental balance tipped in its favor, with corals being under tremendous stress from a world that is changing more rapidly than they are capable of adapting to.

In late 2020, I was hired as the coral biologist and lead of the Coral Response Team for Dry Tortugas National Park, a park that is more than 99 percent of the most beautiful tropical blue water you can imagine. At that time, Dry Tortugas was the only bit of the Florida Reef yet to be infected with SCTLD. Reachable only by boat or seaplane with the nearest civilization seventy miles away in Key West, Dry Tortugas boasted over 30 percent coral cover compared to the rest of Florida's 10 percent cover.[6] Coral cover is a metric for measuring density. Thirty percent cover means that if you looked at a random piece of the ocean floor and calculated how much of it was coral versus sand, algae, rubble, and everything else, 30 percent of it would be live coral. By that measure, Dry Tortugas was the most brilliant gem in the crown of Florida's coral reef.

Reefs are made of corals, which are incredibly cool animals. Individual corals are colonies composed of polyps—little clones that work together to gather food, distribute nutrients, fight competitors, and grow bigger. These polyps are also phenomenal chemists. They pull ingredients from the surrounding seawater to form calcium carbonate, a hard material used to grow their skeletons. And these coral skeletons are what create the foundation of coral reefs.

You might consider corals to be the architects of hustling, bustling, underwater metropolises. Although reefs cover only one percent of our oceans, they support 25 percent of all life in the ocean. Not to mention, it's estimated that one billion people benefit from reefs, which provide food for coastal communities, natural coastline protection against storms, important compounds used in medicine, and tourism that supports economies.[7]

As a coral biologist, I considered myself a steward of these phenomenal ecosystems. Since I was responsible for coordinating fieldwork for the park's coral team, planning and preparing for the arrival of SCTLD was squarely within my jurisdiction. At that time, the disease extended up and down nearly the whole length of the Florida Reef. At its closest proximity to the park, the disease was reported just twenty nautical miles away, near the western edge of the Marquesas island chain. It was inevitable that SCTLD would reach us. Oceanographic models predicted it would arrive at the park's boundaries in March 2021, but we were ready long before then.

We had written the *SCTLD Response Plan*, an extensive and comprehensive document relating all the most relevant research to date and detailing our subsequent short-term response and long-term management strategies for dealing with the oncoming disease.[8] Our plan was run through the bureaucratic tunnel and checked and double-checked by disease experts. We set it in motion, starting with routine reconnaissance surveys, searching every nook and cranny of the park's one-hundred-square-mile waters for the first sign of sickness.

We found it on May 29, 2021, a day I will not soon forget. As I said later in an interview with the *Keys Weekly*, "When we found the first colony . . . with SCTLD, the stark, white lesions of disease, which expose patches of the coral animal's bare skeleton, seemed to glare back at us. SCTLD is distressing to observe, so I felt motivated to do everything possible for that colony."[9]

From then on, we began treating infected corals with an antibiotic paste, a mixture of amoxicillin and Base2B (a special cocoa butter–based formula, created specifically for treating the disease), which was the most effective treatment available. Each day, we mixed the antibiotic treatment by hand or using a KitchenAid mixer, packed it into caulk tubes, and placed the loaded tubes in an icy cooler. Suited up and ready for a series of long dives, with chilled treatment tubes loaded into caulk guns strapped to our hips, our dive team looked partly like sea monsters and partly like strangely outfitted superheroes.

Folks started calling me "the coral doctor." At first, I was proud of the title. (Hubris is perhaps the most rampant vice among scientists.) But as days of diving melded into years, I began to see myself and the work I was doing differently. I was a bird-masked medieval witch doctor, holding a fistful of rose petals under the nose of a bubonic patient, hoping it was a working cure. The dive team joked about being in the business of "saving corals," but as the disease outpaced our efforts, I was humbled by the fact that I could no sooner save a coral's life than I could my own. I was no savior. I was just a marine biologist, trying my best to keep my head underwater (and figuratively above water), while often observing the fading turn from life into death. If I was a doctor, I was working in hospice care.

News articles mostly tell people what they want to hear. My story is of what I saw, from before the disease arrived to the aftermath of its full staying power, becoming endemic in the park. I can speak with the quiet confidence of field experience, studied knowledge, and faith-based perspective that which time will tell. Will the corals make it? Will we scientists?

The Windjammer didn't make it. For months, it had weathered the seasons of SCTLD with hardly a sign of sickness. Then all at once, as if lit with fire from within, coral faces melted off their skeleton. There was more disease than live tissue, and then there was almost no tissue at all. The profound environmental heritage of the site was reduced to a mere handful of surviving colonies within just a few weeks.

The jewellike corals became stone. The damsels and jacks swam aimlessly around the wreck, looking confused and betrayed, as if their neighborhood had been consumed in a fire. The goliath grouper was nowhere to be seen, vanished into the mist. I knew it was only a matter of time before the ecological succession of musical chairs would begin, with organisms competing for niches and migrating to new places—some far from their former community. Perhaps the process had already begun, for certainly the music had stopped.

Thank goodness, Porky was still there. He looked puckered and depressed, waving me over with a flimsy fin. I followed, relieved to see my old friend. His liquid-looking eyes bulged with emotion, gazing in the direction of home. The two enormous brain corals that had built his house were dullish gray with the encroachment of colonizing microalgae. The corals were long dead, and Porky was saying goodbye.

Consumed in a tremor of emotion, I placed my hand on one of the skeletons to steady myself. It was solid and lifeless. I blinked in disbelief, then sobbed into my mask until I couldn't see. *I'm sorry, Porky.*

It was the last time I ever saw that little fish. When our team surfaced from the dive, we were pink-faced and silent. The saltwater droplets around our eyes were not of the sea.

Watching the violent necrotic disintegration of coral colonies I had long loved, I found myself in the front-row seat of a collapsing ecosystem. Vibrant, colorful, and diverse sites were completely transformed into graveyards, filled with white ghosts before fading to heaps of algae-covered stone and rubble. My love for corals cost me, just as C. S. Lewis predicted.

> There is no safe investment. To love at all is to be vulnerable. Love anything, and your heart will certainly

> be wrung and possibly be broken. If you want to make sure of keeping it intact, you must give your heart to no one, not even to an animal. . . . If a man is not uncalculating towards the earthly beloveds whom he has seen, he is none the more likely to be so towards God whom he has not. We shall draw nearer to God, not by trying to avoid the sufferings inherent in all loves, but by accepting them and offering them to Him; throwing away all defensive armour. If our hearts need to be broken, and if He chooses this as the way in which they should break, so be it.[10]

I had finally been given the opportunity to make a difference for corals, only to have them taken from me. I thought my purpose for being a coral biologist was to win the fight for creatures I loved, but I found instead it was my destiny to lovingly help them say goodbye. If I could best love corals by having my heart broken, I would stay alongside them and offer what help I could as they died. If I could best love God's creation by serving it in grief, I would stay with the trouble, weeping and praising the Creator as I went.

The corals themselves convicted me in my complaints. They suffered graciously. They praised their Creator unceasingly. They lived out the entirety of their days purposefully. It was as if, through the suffering leading unto death, they achieved entire sanctification. As environmentalist John Muir once wrote of a suffering dog that accompanied him on a glacial voyage, "there is no estimating the wit and wisdom concealed and latent in our lower fellow mortals until made manifest by profound experiences; for it is through suffering that dogs as well as saints are developed and made perfect."[11]

We grieve the pain, suffering, and death of this present world, the land of the living. And not only us but the world itself grieves, languishes, groans. "We know that all creation has been groaning as in the pains of childbirth right up to the present time" (Romans 8:22). The simile of childbirth is used to describe the labor of all creation, which connotes the Genesis curse in the Garden of Eden, first experienced by Eve and Adam (Genesis 3:16-19, NIV). For Eve, the "pains in childbearing" (labor) were increased; for Adam, it was "through painful toil" (labor) that the earth would produce food for him. The processes of bringing life into the world, by childbearing and cultivating the earth, became marked with pain. Because of humankind's decision to sin against God, all of creation has experienced hard labor as a consequence.

Labor in either form—childbirth or hard physical toil—is tremendously painful. It is in this arduous, suffering state that the consequences of creation's curse occur. However, Christianity does not by any means condone suffering as a fact of life or necessary evil. In his sermon "Christian Hope and Suffering," Timothy Keller references the philosophical work of Dr. Marilyn McCord Adams, who explains how Christianity treats suffering differently from all other philosophies. "The Stoics said accept suffering. The Epicureans said avoid suffering. The aesthetics and the masochists said embrace suffering, for its own sake. But the gospel does not accept suffering. It does not avoid suffering. It does not embrace suffering. It engulfs suffering."[12] Christianity offers a perspective of past, present, and future reality that cannot be achieved by any other philosophy because the Christian gospel contains a promised blessing that overwrites, usurps, and transcends all present pain.

This blessing means that, as childbearing labor results in a newborn baby and overwhelming joy, and as farming labor produces a fruitful harvest and deliciously satisfying food, all earthly labor—whether the groanings of creation or suffering imposed by our vocational work—is never in vain. And labor is not merely an end in itself. Paul's writings to the church in Rome explain the ultimate reason for our suffering. "Now if we are children [of God], then we are heirs—heirs of God and co-heirs with Christ, if indeed we share in his sufferings in order that we may also share in his glory" (Romans 8:17, NIV). This is the reason: If we share in his sufferings, we will share in his glory. For those who know and follow Jesus, this glory is far from a consolation prize; it is a promise. As Keller says, "If the Christian hope is not just a compensation for what we've lost, but a restoration of the world and the life we've always wanted, that changes everything with regard to suffering."[13]

By suffering now as Jesus did during his human life and death, we will also share in his glory, his resurrection and life after death. No greater blessing can be fathomed than to be a child of God and coheir with Jesus, or as Paul states, "What we suffer now is nothing compared to the glory he will reveal to us later" (Romans 8:18). We can consider how we will look back on our present adversity once God restores his creation.

> The new heavens and new earth will make every horrible thing you've ever experienced nothing but a nightmare. . . . And as a nightmare, [it] will do nothing but infinitely, correspondingly increase your future joy in glory in a way that it wouldn't have been increased if you'd never suffered it. And that is the ultimate defeat of evil . . . to say that evil will, in the end, be the servant of your joy. . . . Our momentary affliction achieves an eternal weight of glory beyond all comparison.[14]

Amazingly, this blessing awaits not just humans but all of creation! Every being groans for the eventual blessing through which everything will be restored and made new.[15] All creation labors alongside the children of God, eagerly awaiting "the redemption of our bodies" (Romans 8:23, NIV). For as through Adam the earth was cursed, so through Jesus the earth will be blessed.[16]

> All creation is waiting eagerly for that future day when God will reveal who his children really are. Against its will, all creation was subjected to God's curse. But with eager hope, the creation looks forward to the day when it will join God's children in glorious freedom from death and decay.
>
> ROMANS 8:19-21

Creation labors in expectation of its rebirth, a new life made possible through Jesus, a life void of suffering and death and wholly filled with freedom and glory.

For those who have received God's salvation and have thereby been brought into the freedom of his adopted children, our purpose for laboring is rewritten. "Always give yourselves fully to the work of the Lord, because you know that your labor in the Lord is not in vain" (1 Corinthians 15:58, NIV). We are called to live out our ongoing labor until the day of complete restoration. Through Jesus, God gives labor meaning as a purposeful movement on behalf of the Kingdom of Heaven, even from within the land of the living.

"The land of the living" is a phrase repeated throughout the Bible. In the New International Version, it appears in five different books:

Job, Psalms, Isaiah, Jeremiah, and Ezekiel. The phrase is used to distinguish the present state and status of the created, fallen world from that of the afterlife. For example, "He will uproot you from the land of the living" (Psalm 52:5, NIV) and "I said, 'I will not again see the LORD himself in the land of the living; no longer will I look on my fellow man, or be with those who now dwell in this world'" (Isaiah 38:11, NIV). I believe this same phrase is the most fitting description of where we now find ourselves—isolated to a land where living means suffering that leads toward inevitable death. All creation has been exiled here.

We cry out with longing for a world in which we do not yet find ourselves. We live waiting for the fulfillment of a promise, groaning to be set free from the consequences of sin. It is in this chaotic jumble of beauty and terror—seeing more so every day the reality of our world as broken, fragile, and dying—that we find God.

God's Word reminds us of our exiled existence and future-awaiting hope, or as Rainer Maria Rilke put it: "Nearby is the country they call life."[17] The true country, where we were meant to live, is called the Kingdom of Heaven. We will be there someday. But for now, we are still here.

My heart cries out for something better—for us, for the land of the living in which we roam. To watch ecosystems shipwrecked by the outcomes of human sin and our fallen world—to see the immediate effects of death transforming thousands of strong animals, vibrant with life and pulsing with song, into silent skeletons, brittle as dust—this is our curse. And yet this is also our calling, for we are here to do more than merely watch.

We move through suffering into the pursuit of deeper meanings only God can create. He calls us to keep living, grieving, suffering. To love as wholly and irrevocably as he does. To steward and protect. Yes, we weep. We will weep yet more. The other day, from my living room window, I spotted a cardinal fallen lifeless to the ground. God saw it long before I did, for "are not two sparrows sold for a penny? Yet not one of them will fall to the ground apart from the will of your Father" (Matthew 10:29, BSB). If our Creator sees the fall of even sparrows and cardinals, then certainly he sees corals. He sees us. And nothing can keep him from us, "for I am convinced that neither death nor life, . . . nor anything else in all creation, will be able to separate us from the love of God that is in Christ Jesus our Lord" (Romans 8:38-39, NIV).

In the land of the living, terror and beauty exist side by side. The terror is a reminder of what happened in Eden and the broken remnants of it left on earth; the beauty is a hint of the promise awaiting final fulfillment in the glorious Garden City, the Kingdom of Heaven.[18] We are here laboring amidst both, marveling at God's character written on creation and saying sorrowful goodbyes when called to do so. Because of who God is, we understand that sorrow mixing into beauty yields creation's groanings, waiting to join the promised Kingdom. I remember the Windjammer and continue diving, continue laboring. I repeat the words of the psalmist, "I am confident I will see the LORD's goodness while I am here in the land of the living" (Psalm 27:13).

2

the bothers of beachcombing

Cheerful chirping of nesting birds was suddenly surmounted by the joyous, rushing cacophony of invisible ocean. As my pace quickened, the canopy of trees whispered overhead. Where the forest trail ended was where the beach began, massive waves crashing like happy cymbals in an orchestra of sound and ever-changing colors—my favorite place in the world. And now here, on the cusp of the shore when my legs most wanted to run and my hair sought to escape and my heart longed to dance with the waves, my entry was barred.

As far as the eye could see to either side and some fifty feet ahead lay an immense wreckage of driftwood. Not the small, wood chip–like pieces that can be walked on, but a jungle gym of washed-up trees and ancient, gnarled branches worn smooth as splinters. It was as though Ents from *The Lord of the Rings* had fought here, leaving carnage behind.

Reaching the water required a process of cautiously clambering over the splintery Ents. Sometimes precariously balanced pieces would shift beneath my weight, threatening to pinch or crush.

There is a similar array of wreckage between us and Truth. Just when you think you've arrived, the path is obstructed by a battlefield of doubts, discouragements, and distractions. Climbing over this obstacle course requires work, but you must do it to reach the thing you've searched to find. If you come out on the other side a little worse for wear with a couple of splinters in your hands, you'll have done pretty well. And you'll finally be at your favorite place in the world.

The image of this driftwood zone may be helpful to keep in mind as we move into deeper conversations about the intersection of faith and science. There is much to explore between the two, beginning with an assessment of how faith and science relate. And more specifically, can they coexist?

Faith and science run into some of the same issues. Both are conducted by flawed human beings capable of making notoriously large mistakes. "Our thought and literature remain energized by the widespread belief that all of prehistory and history, including every great transition, somehow served the purpose of placing us upon the Earth," says Harvard sociobiologist Edward O. Wilson. "Everything, it has been argued . . . was meant for us."[1] Notably, "the obsession with putting ourselves at the centre of everything is the bane not only of theologians but also of zoologists."[2] This anthropocentrism, this human centeredness, creates an array of challenges. But arguably, the first issue faith and science run into is that of each other.

There is a conundrum. Science may make claims that faith does not, just as faith can make claims that science cannot. Contrary to what excited answer seekers may think, science is not in the business of "proving" anything. Yes, it asks hard questions. It collects evidence, as much of it as possible. With enough supporting evidence, the "answer" morphs from hypothesis to theory and potentially garners a vast scientific consensus. As such, science is in the business of learning everything yet proving nothing. Therefore, it cannot possibly confirm the existence of God. Science may evidence something at the beginning, something large and difficult to quantify, something beyond us. But science is not designed to contribute to religious or philosophical thought. Morals and ethics may be brought to science, but only insofar as they provide a societally safe framework within which science may conduct its activity. Any biases must be accounted for. Built largely on rationality and objectivism, science can never know with absolute certainty, although it seeks to get as close to it as possible.

To think that science is inferior because of this is to completely misunderstand its purpose and function. Science is like a machine that was built with legs to go from motionless to crawling, crawling to walking, and walking to sprinting. To expect science to prove anything beyond the shadow of a doubt is to demand that this machine sprout wings and fly.

Science simply cannot make the claims that faith can. Where science runs up against a wall, faith floats over like a ghost. "Now faith is the assurance of what we hope for and the certainty of what we do not see" (Hebrews 11:1, BSB). Faith does not get stuck in the cyclic demand of the scientific process, being required to move from observation to hypothesis and then from experiment to conclusion, only to be repeated many times over. Faith means putting our hope and certainty in God, unseen and unquantifiable as he may be. Faith allows God to inform our understanding of the world, ourselves, and everything else. He defines Truth. He *is* Truth.

But faith runs into difficulties when it makes claims that go against the scientific consensus, when what is named Truth by faith does not match what is supported as Truth by science. "By faith we understand that the universe was formed at God's command, so that what is seen was not made out of what was visible" (Hebrews 11:3, NIV). And from the scientific perspective, the idea that "what is seen was not made out of what was visible" is more poetic than prosaic. It ricochets off the ears of science, seeming nonsensical, assuming, and biased, as though a rejection of a structured system reliant on mountains of hard-won evidence. If both faith and science ultimately seek to arrive at the same end of Truth, why do they squabble so?

Most often, bickering starts at the very beginning, over supposed discrepancies of how the natural world originated. Science speaks of the instantaneous physical expansion of the universe at the beginning of time. This phenomenon is called the big bang. Faith describes how God spoke and the universe took shape precisely as he designed. This concept is called creationism. Science claims that enough random molecular collisions could result in the formation of life, given infinite lengths of time to achieve these low probabilities. Faith claims that God spoke again to create vegetation and creatures, into which he breathed the "breath of life." Science strongly evidences that different kinds of animals came from common ancestors, and that time, random mutations, and natural selection shaped different outcomes for the ecological world. This is called evolution. Faith holds to the idea that God made all creatures "according to their kinds" (Genesis 1:25, NIV), indicating specialized, intentional design of both form and function for each living being. This is called special creation.

How are we to reconcile these two accounts? In many ways, they seem opposed, even pitted one against the other. As a Christian and scientist, who both approves science and trusts in the scientific method and yet loves Jesus and believes the Bible, I do not think it is possible to move forward into any sort of exegesis on marine biology until we struggle through this question. We must decide whether we will avoid crossing the driftwood barrier or will struggle our way over it. We can choose to either ignore the question or think critically. The first will lead us nowhere and the second will direct our path toward Truth, but we must choose.

Navigating my way over the beach's driftwood barrier was particularly difficult as a kid. I have memories of long splinters becoming embedded in my palms and knees. I'd surmount one challenging pile of logs just to face off with an enormous tree trunk, its bedraggled roots pointing toward me like a threat. But even among these obstacles, there was the occasional beauty. A feisty crab might keep me company on part of the journey, scaling a piece of driftwood like a spider climbs a wall. I sighted an octopus once, slightly dehydrated and looking a bit worse for wear, but cautiously oozing its return seaward. Even the driftwood itself could be beautiful, with glistening sides worn smooth by sand and salt, some pieces oddly shaped or perfectly pocket-sized, begging to become a personal treasure. There was purpose in my journey over the driftwood.

Likewise, there is value to asking hard questions, as scientists do. And there is value to wrestling with doubts, like the faithful. We cannot have "the assurance of what we hope for and the certainty of what we do not see" (Hebrews 11:1, BSB) if we do not, on some level, acknowledge that we trust in eyesight we do not currently possess. We are desperately hoping and cannot see, yet we are choosing to go out on a limb and run the experiment anyway. When we ignore questions and doubts, we hand them great potential to drift us off into oblivion. The second we forego curiosity and the willingness to ask questions, scientific credibility goes up in smoke. We pile the kindling and pour gasoline on the driftwood burn-pile that was once our faith, strike the match, and wonder why it goes up in flames. To prevent doing so, we must ask the hard questions, face the doubts, and clamber over the driftwood. We must design our experiments and let our faith be tested. Otherwise, neither science nor faith are worth having.

But are both science and faith truly worth having? And are they compatible with each other or are they mutually exclusive? We need to dig a little deeper into one of the core doctrines of Christianity in order to set the stage for this discussion. Core doctrine is basically the "essentials" of faith—the things that a person must absolutely believe in order to be considered a Christian. Perhaps core doctrine is best summed up by the Nicene Creed, which includes a specific statement about how everything was created: "I believe in one God the Father Almighty; Maker of heaven and earth, and of all things visible and invisible."[3] In other words, God created all things—a concept which, as previously mentioned, is called creationism. As core doctrine, creationism is a view upon which all Christians must agree.[4]

However, Christians can have differing views on how creationism actually happens. Considering these views is useful because they provide multiple options for how faith and science may fit together. Because creationism is core doctrine, these four perspectives of how creationism works are named for their relationship with it. All four views exist on a spectrum, where the first is the most well-known and difficult to reconcile with modern science and the last is lesser known and most easily reconciled. These views include young earth creationism, old earth creationism, progressive creationism, and evolutionary creationism.

Historically, young earth creationism may be the most widely preached version of the biblical creation story. Also known as seven-day creationism, it holds to a direct, literal interpretation of the creation story in Genesis 1. Based on the English translation of the original Hebrew, young earth creationism asserts that the world was created by God in exactly seven days. He created the universe and living things in six literal twenty-four-hour days and set aside one final day of rest. Furthermore, young earth creationism uses a literal interpretation of biblical genealogies to predict the earth's age as 6,000 to 10,000 years old.[5]

While old earth creationism uses a direct, literal interpretation of Genesis 1 as well, it syncs with scientific estimates of the earth being 4.5 billion years old. Old earth creationism often draws upon gap theory, the belief that an unspecified length of time (a "gap") occurred between the first two verses of Genesis. "In the beginning God created the heavens and the earth" (Genesis 1:1). Then an indeterminate period of time began, extending perhaps billions of years. The second verse describes this period: "The earth was formless and empty, and darkness covered the deep waters. And the Spirit of God was hovering over the surface of the waters" (Genesis 1:2). The hypothesized gap between these verses would account for the physical and geological processes theorized by science, while still allowing for a literal interpretation of Genesis.[6]

Progressive creationism, a theory that first emerged a few centuries ago, has also traditionally fit under the umbrella of old earth creationism. Unlike its predecessors, however, progressive creationism does not require a literal interpretation of Genesis 1. This view asserts that the Hebrew word for *days* in the Genesis creation account is better translated as "periods" or "ages."[7] Therefore, each of the six days represents an extended epoch rather than a literal twenty-four hours, presuming that God created over very long periods of time. Because of this, progressive creationism is in accord with scientific estimates of the earth's age.

The fourth theory, evolutionary creationism, also referred to as theistic evolution, primarily relies on a functional or allegorical interpretation of Genesis 1. Rather than reading this chapter for the purpose of scientific cosmology, evolutionary creationism sees Genesis 1 as describing the story of God's creativity, intentionality, and care in creating the universe, thereby revealing his character. This view contends that God used evolutionary processes to bring about diversity of life on earth. As such, God's creating actions were not limited to six "day" periods of time but extend even up to and beyond our current time. In this view, God is still actively at work, creating more of his world as new creatures arise and are discovered by modern science.[8]

As a reminder, a person may hold any of these four views and still be a Christian. Creationism itself is the core doctrine, but our view of how it functions includes room for interpretation, as it is a secondary issue to salvation. Belief in a particular option is not required to maintain faith. So long as science is in accord with creationism, there is no reason why faith and science shouldn't join hands along their pursuit of Truth.

Free of the driftwood, I was sprinting through the surf, zigzagging between globs of bubbly sea-foam that gathered together like dust bunnies. To my left, the ocean roared currents and laughed ripples at my feet. To my right, the coastline swept upward in a frenzy of piled rock, etched above the high tide line with the hugging roots of evergreens. Ahead, large boulders the size of cars rose as blurry shapes through the drifting mist, tendrils of fast-moving vapor curling around them. I ran toward the boulders and joyfully weaved my way between them.

The sand was softer here. Where my feet had padded hard on the edge of surf that kissed the beach, here they sank several inches. It was like running on a layer of pudding. Curving by the side of a boulder, the sand became dangerously gooey. Pudding-like muck sucked at my legs and suddenly, with a comical *slurp*, I was trapped waist-deep. I heard the ocean laughing in the background, but my view of it was blocked by the giant boulder I had sunk beside. After several moments of intense struggle, I gave up. Like Christian trapped in the Slough of Despond, I was absolutely stuck.

Thankfully, I had been running with a friend at the time. After he recovered from laughing at my unfortunate but comical situation, he grabbed my arms and wrenched me free. I'll never forget the regretful *slurp* sound that the pudding sand made as I was finally removed from its suctioning grasp.

Learning new things is a bit like recognizing you are stuck in gooey sand. You're waist-deep in your own thoughts, opinions, and perspective. You probably can't see the entire landscape, even though it's all around you. And on your own, it's extremely hard to get out. But free from the sand, you can run around again. You can sprint through ocean surf, hearing and seeing it gurgle, while weaving between boulders to chase gulls beyond the rocks. You can even realize that there are different kinds of sand.

In life, it's a challenge not to get stuck in one set way of thinking. The sand wants to suck us in. But we have the opportunity to scour the entire beach. To consider other perspectives. To run as free as a fiddler crab between the low and high tide lines.

I hold nothing against my friend who laughed at me before pulling me free. But metaphorically speaking, we ought to be charitable toward others when they are stuck. We should regard one another leniently and favorably, kindly offering a hand. Just because we have different perspectives doesn't mean we can't help each other. Just because we disagree doesn't mean we shouldn't be willing to hear the other side. Maybe they're onto something. Maybe, in fact, they're the ones freely roaming the beach while we're stuck waist-deep in pudding sand.

To keep from getting mired down in a particular view of creationism, we can consider the strengths and weaknesses of each perspective. Among the four views, initial differences lie in two areas: how the Bible is read and how genealogies are used. There is nuance to each. When reading any important text, we must read it as it was meant to be read—as the author intended without projecting preconceived notions or opinions onto the text. We are endeavoring to discern Truth. So what does the Bible truly say? How do we read it rightly?

There is debate about whether certain parts of the Bible should be read literally or allegorically. Literal reading is where words on the page are expected to mean what they say on a practical level. If the sky is described as gray, a literal reading of the text would determine that the sky was the color gray. Allegorical reading is where the words on the page are expected to contain layers of meaning between the lines. If the sky is described as gray, an allegorical reading of the text might lead someone to conclude that the sky is covered in clouds, giving off a melancholy feeling. While literal and allegorical reading can lead to dramatically different interpretations, both come into play in a fascinating way depending on the kind of writing.

The Bible contains three main kinds of writing: narrative, poetry, and prose discourse. Narrative writing includes history, biographies, and parables, which compose 43 percent of the Bible. Exodus and Matthew are examples. Notably, narrative writing is best understood through literal reading. In contrast, poetry includes creative language, imagery, and metaphors to express emotion and wisdom. Thirty-three percent of the Bible is poetry, including books like Psalms and Ecclesiastes. Poetry is best understood through allegorical reading. And lastly, prose discourse refers to analytic discussions such as speeches, letters, and essays, which compose 24 percent of the Bible. Romans and 1 and 2 Corinthians are examples. Prose discourse is best understood through a combination of literal and allegorical reading.[9]

By considering the different kinds of writing present in the Bible, we can see that both literal and allegorical reading are required for complete understanding. In short, we need to be able to read literally and allegorically to tease out the true meaning of the text. And this also applies to how we consider genealogies, which are records of descendants from a common ancestor. Each one is a family tree.

Genealogies can be a major sticking point when it comes to reconciling the Bible's creation story with modern science, so it's helpful to consider what genealogies are, how they work, and how they relate to the creation story. The Genesis creation account is followed by a genealogy of Adam (in Genesis 5), which is used specifically within young earth creationism to calculate an age of earth that does not fit with scientific estimates.[10] What are we to do with this genealogy? And more specifically, is it reliable for interpreting the age of the earth?

While genealogies are relevant in some scientific studies, they have a special purpose in the Bible. Biblical genealogies, called *toledot* in ancient Hebrew, are different from how we think about genealogies today. They contain deeper meanings and were written for theological purposes rather than mere historical accounting.[11] In other words, genealogies in Scripture were written to help make connections between biblical stories as part of God's grand epic. They provide a framework and reference points for the story that follows. For example, the genealogy of Adam in Genesis 5 bridges the gap between the story of the Fall (when sin enters the world through Adam in Genesis 3) and sets the stage for the Flood (when Noah and his ark come on the scene in Genesis 6).[12]

In addition, there are two forms of biblical genealogies: linear and segmented. Linear genealogies are like tree trunks. They have depth, showing vertical relationships (such as father to son). Comparably, segmented genealogies are like branches and leaves. They have breadth, showing horizontal relationships (such as husband to wife).[13] Scholars agree that both linear and segmented genealogies can be described as fluid.

"Fluid" means that the same genealogy could have been recorded several different ways, depending on its purpose.[14] If a genealogy was written to precede a political event, it would be written very differently than if it was intended to provide commentary on a social occurrence. The genealogy's function determines its form. The term *fluid* also refers to the fact that in Hebrew, "son of" does not always mean that the person was the literal son of another person, but rather that he was a descendant.[15] In the same way, "father" does not always mean the literal father of the person but rather an ancestor. In many cases, scholars have noted that genealogies even omit names and skip generations![16] This is because genealogies are frameworks for understanding the stories they precede. They are not exhaustive lists of everyone who existed within a family tree.

Genealogies are reliable only so long as they are read for their intended purpose. They are unreliable in determining scientific information because they were not written to help us know the age of the earth; they were written to help us know the Creator of the earth.

Most biblical scholars and theologians agree that the creation account in Genesis was never intended to provide a scientific or historical chronology. Although Genesis recounts a history of sorts, its author was likely writing the first chapter in the format of a traditional poem, which functioned as a prologue. The material was repeated in more prosaic language in the second chapter.[17] This provides an explanation for the repeated phrase throughout the creation account, "And there was evening, and there was morning . . ." (Genesis 1:5, 8, 13, 19, 23, and 31, NIV). This repetition is likely a lyrical device within the poem.

Knowing that Genesis 1 was written as poetry should enlighten and guide how we read it. While much of Genesis can be read literally, chapter 1 is probably best understood through allegorical reading. And since Genesis was not written for the purpose of explaining cosmology beyond the core doctrine of creationism, the possibility of a gap between the first and second verses (as proposed by old earth creationism) is likely not intended on the part of the author.[18] However, Genesis 1 does provide a theological framework, as expressed by all four views of creationism, within which we can learn to better understand creation and the Creator.

When consulting the Bible, we ought to focus on literary structure and theological themes rather than imposing scientific conclusions that may or may not be relevant, intended, and ultimately true.[19] The Bible is not a science textbook. So why do we expect it to tell us exactly how the universe came to be? That has never been its aim. Rather, the goal of the Bible is to tell the story of God from beginning to end. Regardless of where we stand on our reading of the Genesis creation account, the world is no less miraculous and wonderful for having been created by God in a surprising way.

As we consider the different views of creationism, we begin to see that faith and science are not mutually exclusive. Any reason for their supposed incompatibility is due to a misunderstanding, either of the Bible and how it should be read or of science, which is ever-changing and advancing. Although both faith and science function differently, they both shoot for the common goal of Truth. As Pope Francis has written, "Science and religion, with their distinctive approaches to understanding reality, can enter into an intense dialogue fruitful for both."[20]

Furthermore, Chuck Colson notes that "faith requires no surrender of the intellect."[21] You do not need to shove aside doubts or refrain from asking questions. You do not need to "switch off your brain" in order to have faith. On the contrary, turn on your critical thinking! Let us clamber over the driftwood barrier, asking hard questions and engaging our intellects, facing the bothers of spiritual beachcombing. For if both faith and science are sincerely seeking Truth, they will ultimately arrive at the same end: God himself.

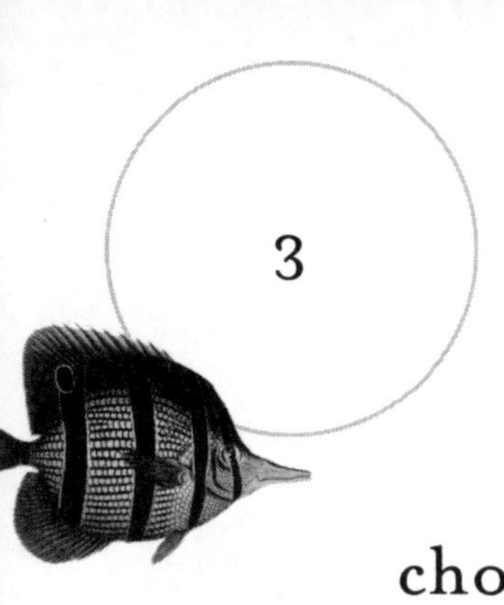

3

choir of creation

Wrapped in blue, flurries of silver particulates float gracefully in and out of view. I hear before seeing—a quiet picking, barely there, like the gentle plucking of a guitar string, dulled against the grinding backdrop of other curious harmonies. Scratchings, scrapings, comings, goings. A flurried whoosh as a tail kicks up a miniature billow of soft sand. The gliding swish of scales against scales. The crackling of a cleaner shrimp nestling into the rocky fortress it calls home. The crisp clicking of amberjacks with brilliant blue eyes, beckoning friends to join the school. The squirting jet of an opalescent purple anemone, gel-like and luxurious against rocky outcroppings. Miniscule reverberations of thousands of coral polyps extending and withdrawing, catching plankton with their sticky tentacles. The bass-like *whoomph* of a defensive red grouper guarding its territory, and the near silent scraping of an arrow crab's claw as it combs a manicured algae garden. The sharp crackle of hungry parrotfish teeth against coral skeleton. This reef is alive with noise, the splendorous cacophony of an ecological orchestra.

I listen as a careful witness to this blend of intertwined musicality expressed through sound and motion and color, agreeing with the musings of Philippe Cousteau Jr. when he reflected on a dive in the Red Sea:

> "If there is a *Garden of Eden* on earth . . . this must be it," I thought to myself. . . . Every color imaginable could be seen stretching out along the contours of the reef. Fish of every shape and size, from gobies no larger than the head of a pencil to enormous groupers and shimmering reef sharks, darted all around me while enormous sea fans and sponges dotted the landscape.[1]

As a marine biologist, observing these sights is part of my job. As a Christian, listening to the reef's song is my calling. The scuba equipment I wear is far more than the acronym for "self-contained underwater breathing apparatus" implies. It isn't just a uniform; it's my admission ticket to the front row of the most wonderous music my ears will ever hear.

Reef noises in all their glorious variety are indicative of biodiversity, a measure of how many different kinds of organisms live within an environment. Biodiversity may be compared to an orchestra or choir, where individuals form different parts of one unified song. While the melody alone may be pretty with a single, strong voice, the song is enhanced to a greater scale of beauty when that voice is accompanied by many—when alto, tenor, and baritone parts join the mix.

The same can be said for the role of biodiversity in ecology, defined as the study of relationships between living things and their environments. Connecting relationships compose a larger network, called an ecosystem. One living creature may make an impact and contribute to the functioning of its environment, but many creatures of different kinds contribute far more, creating what is typically considered a healthier and more stable ecosystem.

Biodiversity, then, values differences between species. The ability of the created world to perform different functions is what lends to the earth's amazing and complex systems and processes. It is essential to note that these differences are prized not for the distinctions and divisiveness they may create, but rather for the strength that comes from unified variety.

Each species plays a unique ecological role, filling an environmental niche that only it can, as though singing a unique part among many voices. These independent voices, when combined in the form of an ecosystem, create a thing of stability and strength, just as the meeting of many voices forms a choir that swells to create beauty beyond the reach of one singular voice.

When fewer individuals sing a part, that part becomes quieter. This is what I encountered at the Windjammer when corals fell silent in the wake of disease. When only a few individuals of a species remain, the functional role of that species (the job they perform within their niche) can be lost. While the species may still be represented in that ecosystem, its voice within the choir cannot be distinguished from the rest, just as if nobody were singing that part. When this happens, the species may be considered "functionally extinct."

Some scientists argue that functional extinction is hardly any better than true extinction, and indeed, from an ecological perspective, the remaining individuals have little to offer besides their inherent value as living things. But when organisms go extinct, or when a local extinction occurs (where an organism is described as going "extant"), the overall biodiversity of that ecosystem is diminished.

When biodiversity continues to decrease, new problems can arise. Stripping biodiversity destroys the subtle equilibrium of complex relationships within environments. Simply put, the loss of a species can mean the loss of a landscape, like I saw firsthand with the death of Porky's home. The loss of many corals can equate to the loss of a reef system. With the death of this one type of animal, hundreds to thousands of other organisms (animals, plants, and microbes) are impacted. The damselfish that lives inside branching coral must find a new home. The parrotfish that munches on hard corals will have to travel to find a new snack. The zooxanthellae algae that live in harmonious symbiosis with corals may potentially die.

While biodiversity is important for these biological reasons, there is spiritual value to biodiversity as well. This is because "God dwells in His creation and is everywhere indivisibly present in all His works."[2] Much like the perspective and creativity of an artist is revealed through their art, the character of God can be revealed through everything he has made. One catechism puts it like this: "Each creature possesses its own particular goodness and perfection. . . . Each of the various creatures, willed in its own being, reflects in its own way a ray of God's infinite wisdom and goodness."[3] Microscopic tardigrades, capable of thriving in even the most extreme environments, demonstrate God's tenacity and attention to detail. Stars shining into our telescopes from faraway galaxies display his vastness and majesty. Relational, nurturing kangaroo mothers embody his tenderness and affection. The sleek silhouette and countershading of reef sharks symbolize his agility and ingenuity. Maverick waves surging and breaking in an explosion of water communicate his power and strength. Mobile, shell-recycling hermit crabs reveal his innovation and imaginativeness. Each solitary planted seed hints at his covenantal nature and care for our futures. And gooey, big-nosed, deep-sea blobfish perhaps reveal his sense of humor.

Effectively, nature talks to every human about God, "since what may be known about God is plain to them. . . . For since the creation of the world God's invisible qualities—his eternal power and divine nature—have been clearly seen, being understood from what has been made, so that people are without excuse" (Romans 1:19-20, NIV). In other words, God's character is so deeply integrated in creation that humans are "without excuse" in their decision to accept or reject God. If only we take time to see it, creation makes God's character clear. If we do not see it, that is because we choose not to look. These verses also highlight that God's "eternal power and divine nature" are revealed through the created world. What else can we learn of God's character through creation? Inquiring scientific minds will crave greater specifics.

Surely, there are countless things to learn, but as an ecologist, I would argue that a love for biodiversity is toward the top of the list. If biodiversity is valuable biologically, and God created both diversity and biology, then surely God values biodiversity. God designed the principle that biodiversity strengthens ecological communities, just as diversity strengthens human communities. In short, it is a central tenet to our understanding of ecology without which we would not comprehend all the phenomenal interconnectedness that God designed in creation. Creation was made for relationship with itself, with humans, and with God. "The community of the centre of all creation," says Scottish author George MacDonald, "suggests an interradiating connection and dependence of the parts."[4] All creatures were designed to be in integrated ecological relationships with different living things. The meeting and meshing of these different creatures in relationship with each other is the core of biodiversity.

God's delight in diversity is also reflected in the biblical metaphor of the unified church being the body of Christ. The apostle Paul speaks about how members of the church should eagerly desire spiritual gifts that will enable them to perform unique functions for the benefit of all and to the glory of Christ (1 Corinthians 12:1-11, 27-31). Within this context, he compares members of the church to members of the human body, such as hands, feet, and eyes. He reminds us that a body, although made of many parts, works together as a unit; in the same way, the church, although it is made of many individuals, works together in love as a unit—which is why it's called the body of Christ.

> If the whole body were an eye, how would you hear? Or if your whole body were an ear, how would you smell anything? But our bodies have many parts, and God has put each part just where he wants it. How strange a body would be if it had only one part! Yes, there are many parts, but only one body. . . . If one part suffers, all the parts suffer with it, and if one part is honored, all the parts are glad.
>
> 1 CORINTHIANS 12:17-20, 26

God created biodiversity, where species exist within an ecosystem, for a similar purpose. As the church can be compared to body parts that function together as a unit, aspects of creation work together to form functional ecosystems. If the whole ecosystem were simply one kind of thing, where would the biodiversity be? If the whole ecosystem were algae, where would the means of predation be? If the whole ecosystem were a fish, where would the means of food be? But in fact, God has arranged the parts of an ecosystem, every species and individual, just as he wanted them. If they were all one species, where would the ecosystem be? As it is, there are many species within one ecosystem. If one species suffers, the whole ecosystem suffers with it.

Just as many voices build a choir, many parts form a body. Take away one piece, and you take away from the quality—the health, function, beauty, and ability to praise God—of the whole. This speaks to the inherent and functional value of living things, the worth attributed to them by existing and functioning in their ecological role, by filling their particular niche. More specifically, it speaks to the doxological value of living things, having been created "good" for God's glory.

I first heard the singing in college. I was working at a field station in the San Juan Islands off Washington State, going for daily hikes to collect insects, record songbirds, and core trees.[5] I was alone and stopped to sit along the eastern edge of Spencer Lake, where the verdant ground rose up in massive marshmallow-like lumps of moss. Overhead branches rustled and swayed in the feathery breeze. Reeds quivered along the shore while lily pads shifted ever so slightly as dragonflies alighted and bounced to and fro. A fish's mouth momentarily touched the water's surface, sending ringlets of ripples. A hawk soared overhead, silent and full of glory against the valiant blue sky. My fingers felt the glossy yellow buttercup petals that sprang up amidst the plushness of the vibrant moss. And then I heard it: singing. Soft at first, to my untrained ear. But undeniable. It was as though I was surrounded by a surreal sort of choir, gently singing indescribable words at the hushed level of *pianissimo*.

I returned to that spot many times in the following weeks, trying to identify the source. Having spent many years striving to see and understand ecology, I had immersed myself in the study of searching for the functions and purposes of nature, as told by science. And now I was standing in the midst of it, surrounded by trees, lakes, fields, and waterways of the world. I let myself take it all in, alert to the sounds but also to the sights and sensations. I watched the golden sun thrill the sky with color every evening before vanishing over that unreachable horizon. I caught bats in nets and felt their racing heartbeats against my gloved palms before releasing them again into the night. I listened to eagles cry victory while bringing food to their young. I fell asleep with beetles and butterflies in thickets of ferns, fronds falling softly over my face. I raced like a wave down pebbled beaches, chasing after otters beyond the rocks. I stood with families of deer as they fed along streams before trotting through dewy meadows. I watched petals of morning glories unfold like the twirling skirts of ballroom dancers, only to refold again in shadow.

Then one day as I overlooked an expanse of green islands shimmering against the dark sea, what had begun as silence blossomed into melodic whispering, which turned into music. The music became many voices. And those many voices became a symphony, a sort of choir. The most beautiful choir that has ever been. And somehow, I found myself surrounded by a song of blazing beauty in which I longed to participate and to belong.

If there had been only a few vocalists, the song, although pretty, would have likely been too quiet to be heard. Had the song's parts been all the same, it would have been unremarkable, too monotonous for noticing. If a bird sings by itself, it might sing with all its might and yet not be heard beyond the reach of a field. If a flock joins it, it may be heard farther, but only as a cacophony. And yet the sound of a flock of birds alighting on branches of budding trees to make and fill their nests, while bullfrogs croak beyond the rustling grass where rabbits feed as clouds scuttle across the sky—here is biodiversity in all its fullness. Here is ecology. And to the trained ear, here is a song greater than that of merely birds.

The song becomes more evident once you know the most central truth of the biblical creation account: that God made creation to be his home—a place where he would dwell. As God declares in Job 38:4, "Where were you when I laid the earth's foundation? Tell me, if you understand" (NIV). Realizing this powerful truth about creation's intended design catalyzes understanding of why there is suffering in the world and why the calling of a coral biologist means so much more than merely caring for corals.

"The most central truth to the creation account is that this world is a place for God's presence," writes theologian John H. Walton. "Though all of the functions are anthropocentric, meeting the needs of humanity, the cosmic temple is theocentric, with God's presence serving as the defining element of existence."[6] In other words, the world was not created for humans to be at the center of all things (although we often live that way). The world is theocentric, with God at the center. The world was created by and for him. Simply put, the world is God's temple. He created functions within it to support the needs of people, but he created everything, including people, for himself. This was his original design.

The theological concept of creation being God's temple has been discussed by numerous highly qualified theologians and biblical scholars. Walton, who wrote *The Lost World of Genesis One*, details God's original design for his world using a series of mounting propositions supported by a slew of evidence:

1. In the Bible and in the ancient Near East the temple is viewed as a microcosm.
2. The temple is designed with the imagery of the cosmos.
3. The temple is related to the functions of the cosmos.
4. The creation of the temple is parallel to the creation of the cosmos.
5. In the Bible the cosmos can be viewed as a temple.[7]

In the beginning, when God created the heavens and earth, day and night, land and sea, and all living creatures, he finished each act of creation by affirming its goodness.[8] "God looked over all he had made, and he saw that it was very good!" (Genesis 1:31). His design was perfect. When the Artist stepped back from his artistry, he saw that it was beautiful, representative of himself, and right.

But unfortunately, while God originally made everything good, all is good no longer. We each know from our own experience that the world is far from perfect. For one, there is death. Extinctions happen. There is also pain, grief, and suffering of many kinds. Creation has been ravaged by the effects of cosmic treason. Human choice led to rebellion against God (called sin) when people chose to reject rather than embrace their Creator.

The perfection that once existed in God's original design and achieved intention has been irreparably maimed. And God, being perfect himself, is no longer accurately represented by his creation. It is as if, after an artist had completed his painted masterpiece, a disobedient child came along and splashed black paint all over it. Signs of the old artistry remain, bits of beauty peeking from between splashes of the black, but the original art is damaged, the intended image no longer clear. This is what has happened to humankind, as well as the rest of creation. Lay theologian and writer C. S. Lewis said: "It is enough to say here that Nature, like us but in her different way, is much alienated from her Creator, though in her, as in us, gleams of the old beauty remain."[9] And so, as we may derive bits and pieces of the character of God from what he created, not all is an accurate representation. We must discern what of the "old beauty" remains and what is black paint.

Creation was designed to point its figurative finger at its Creator. Now, creation points two fingers in opposing directions, one toward its Creator and the other toward sin. Biodiversity can result in either unity or divisiveness. One is of the Creator; the other is not. Because of sin, there is only so much we can learn of God through what he has made. C. S. Lewis, in his book *The Four Loves*, put it like this:

> Nature never taught me that there exists a God of glory and of infinite majesty. I had to learn that in other ways. But nature gave the word *glory* a meaning for me. I still do not know where else I could have found one. I do not see how the "fear" of God could have ever meant to me anything but the lowest prudential efforts to be safe, if I had never seen certain ominous ravines and unapproachable crags. And if nature had never awakened certain longings in me, huge areas of what I can now mean by the "love" of God would never, so far

> as I can see, have existed. . . . Nature will not verify any theological or metaphysical proposition . . . ; she will help to show what it means.[10]

The natural world is no longer a perfect illustration of the one who created it. Creation is no longer an accurate reflection of its Creator. But creation reveals enough of God that we can come to know hints and glints of who he is through it. We must pursue God himself to learn discernment between black paint splashes and original brushstrokes. And as we do, let us consider the original design of creation, those gleams of the old beauty that remain, sometimes as subtle as a song.

We can learn more about the created world and its meaning as God's temple by studying the human-built Temple described in the Bible. The Tabernacle built by the Israelites in the wilderness was upgraded by the Temple that Solomon constructed in Jerusalem. Just as that Temple was tragically corrupted and eventually destroyed, so the temple of creation has been corrupted by the taint of sin. Eventually, creation will be destroyed too. The old heaven and earth will disappear, replaced by a new heaven and earth.[11] But remembering the purpose of the Temple helps resolve any ecological issues we may have with this apocalyptic future, because God promises to restore all things. In the meantime, the earthly temple, much like us and every other creature, is in exile. It was never meant to be this way. The temple of creation was originally designed to perfectly reflect and glorify God; now it does so only in part.

There were many regulations involving the Tabernacle and Temple, the place where God dwelled among people. The Old Testament explains that only specific, designated individuals were allowed to approach God. A series of strict cleansing rituals and procedures were required prior to entering, and the consequences of breaking these regulations was instant death, as evidenced by the various times this unfortunate situation played out in the biblical story.[12]

If this level of care was required before entering a physical, man-made Temple to experience the presence of God, perhaps we should evaluate the implications of God designing the world to be his original temple. Since God created the world and saw it was "good," since he instilled so many hints of his own character into his handiwork, like the signature and fingerprints of an artist on a masterpiece, what does that teach us? How do we treat his creation? How does our treatment of creation reflect our attitudes toward him? How do we enter his temple?

In biblical times, priests were specially appointed individuals whose livelihoods came from serving within the Temple. These people enjoyed the blessing of dwelling in the literal presence of God, having the closest physical proximity to him of anyone. The Old Testament details the ritualistic and sacrificial procedures performed by the priests within the Temple, but the New Testament is clear that these laws were fulfilled through Jesus. The need for ritualistic and sacrificial procedures no longer applies because Jesus came as the perfect, ultimate sacrifice on our behalf.

If the world was designed to be the temple and humans were created to serve within the temple, our functional role in the created world is to serve as priests to God. As the apostle Peter writes,

> You are living stones that God is building into his
> spiritual temple. What's more, you are his holy priests.
> Through the mediation of Jesus Christ, you offer spiritual
> sacrifices that please God. . . . You are a chosen people.
> You are royal priests, a holy nation, God's very own

> possession. As a result, you can show others the goodness of God, for he called you out of the darkness into his wonderful light.
>
> 1 PETER 2:5, 9

Notably, these words were spoken to the early church, not the nation of Israel, evidencing that indeed the priesthood has passed from the appointed Israelites to anyone who is a follower of Jesus. The emphasis is no longer a religious building, a list of regulations, or a particular people group; access to intimacy with God is entirely through Jesus, who makes us priests within creation.

The priesthood was both an honor and a task, and it still is. "Do not be negligent now, for the LORD has chosen you to stand before him and serve him, to minister before him and to burn incense" (2 Chronicles 29:11, NIV). A priest's purpose was to serve God, minister before him, and burn incense. While this may seem a cryptic and ritualistic habit that has otherwise been demolished by Jesus' fulfillment of the Law, we may interpret the burning of incense in a less-literal light after considering the rest of Scripture. Specifically, the burning of incense is biblically associated with the lifting up of prayers. "Accept my prayer as incense offered to you, and my upraised hands as an evening offering" (Psalm 141:2). We can describe the role of our holy priesthood as serving God, ministering before him, and praying. Of course, all this takes place within creation, his temple.

In addition to the symbolic meaning of burning incense as prayer, we might also consider the meaning of the charge to "serve" God and "minister before him." Serving and ministering to God within creation yields a heightened understanding of creation and, therefore, of its Creator. This implies a journey, a process of growing in relationship with God. And as we come to know God ever deeper, learning to see his fingerprints and signature on all the canvas of creation, we will be led to praise him. Throughout the Old Testament, the priests led the people in singing "Give thanks to the LORD, for he is good! His faithful love endures forever."[13] Even now, serving and ministering to God will beget praise.

A primary functional role of priests in the temple, of humans in creation, is praising God, the one by whom "all things were created: things in heaven and on earth, visible and invisible . . . all things have been created through him and for him. He is before all things, and in him all things hold together" (Colossians 1:16-17, NIV). We are to lead the rest of creation to glorify him with us; in fact, creation is already praising him, often better than we humans do!

These nonhuman creatures, though tainted by the effects of human sin, never made the sinful choice themselves. They have not rejected their Creator. Perhaps they are better at remembering him than we are. Because of their consistency in glorifying him with their humble lives, we have better access to witnessing God's character through them and, as C. S. Lewis remarked in *The Four Loves*, learning the meaning of the words *glory*, *fear*, and *love.* Realizing our connection to creation reminds us of our Creator. "When we can see God reflected in all that exists," said Pope Francis, "our hearts are moved to praise the Lord for all his creatures and to worship him in union with them."[14] They remind us that just as they sing solely for him, we ought to as well. They teach us that we were called to be a holy priesthood, directing the praises of the temple to the one who resides there. Simply put, humans were designed to be the choir directors of creation's praises to God.

This is the vocation of creation. And the tremendous biodiversity of creation lends rich harmonies to this choir's praise of our Creator. The greater the ecological diversity, the greater potential to maximize praise. And leading this praise is the calling of humankind—particularly God's followers, his priests on earth—that we might urge all things to

> Worship the LORD in all his holy splendor.
>   Let all the earth tremble before him. . . .
> Let the heavens be glad, and the earth rejoice!
>   Let the sea and everything in it shout his praise! . . .
>   Let the trees of the forest sing for joy
> before the LORD.
>
> PSALM 96:9, 11-13

As we walk through the forests, may we interact with the world like Saint Francis of Assisi (whom some have dubbed "the patron saint of ecology"), urging the trees and sky and birds to glory in their Creator and sing of his character their whole lives long. As we scuba dive in the ocean, may we sing as we swim along with all the creatures of the deep: "Praise the LORD from the earth, you creatures of the ocean depths" (Psalm 148:7). As we go through life, might we always encourage, "Let everything that breathes sing praises to the LORD!" (Psalm 150:6).

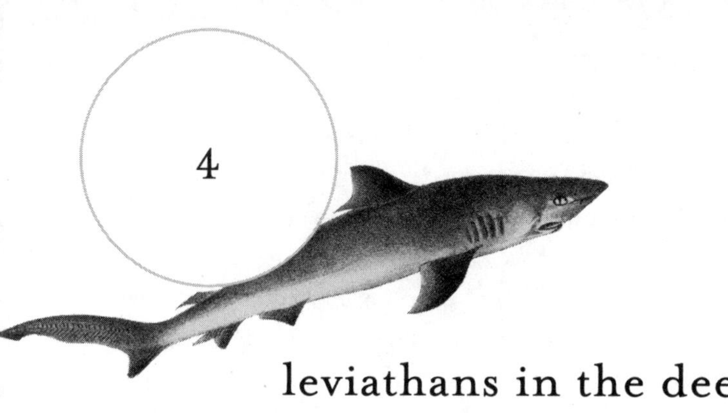

4

leviathans in the deep

"Do you swim with sharks?"

It's the first question most people ask when they discover my professional background. Sharks are known as the villainous stars of ocean-themed horror movies. I never saw *Jaws* because I already possessed an intrinsic, deep-rooted fear of these predatory fish with their multilayered rows of teeth. My earliest childhood memories are of this recurring nightmare: A shark appears. It sees me and begins to follow, slowly at first. Then fast, too fast. Its jaws open. I can't outswim it, and right before the monster chomps down, I wake up in a cold sweat.

It was an inconvenient fear for a budding marine biologist. I kept it to myself. But the shakily asked question, "What if there's a shark?" lurked in the corners of my mind every time I touched water. Even as an adult in a hotel swimming pool, the thought would flitter through my mind. Fleeting, but there. The nightmare haunted me.

Anyone scared of sharks is in good company. Since ancient times, people have feared sea monsters. Horrifying legends abound. Artists rendered dangerous, mythical creatures in all their wretched and glorious detail. Maps of the known world frequently included illustrations of monsters residing in various pockets of uncharted sea, evidence of how fear taps into the imagination's most creative resources. But maybe it wasn't all the stuff of dreams.

Legends are often based on real figures. Myths contain seeds of truth. Perhaps in days not so long ago, before humankind tamed and exterminated much of the world's megafauna, these monsters existed. Shadowy creatures with rings of razor-edged teeth chased down any boat plucking its way across the sea, and terrible yet beautiful half-fish, half-women were either a true threat or a veritable aid. Indeed, the myth of sirens—those seductive half-woman, half-bird creatures who used song to attract and then attack sailors—is represented across a diverse array of cultures and geographic regions, including ancient Assyria, Greece, Asia, Africa, and Europe. In many places, these sirens became interchangeable with mermaids. So maybe there really was something in the water back then. Or maybe people just knew less; we like to believe we are wiser now. Nessie was just a figment in the Scots' imagination, right? And yet just a few years ago, paleontologists found the fossilized remains of a roughly thirty-three-foot-long sea dragon, an ichthyosaur, which is said to have frolicked in the oceans some 180 million years ago.[1] Sea monsters really did exist.

Besides references made by other ancient literature, the Bible refers to one particular sea monster by name, as if the cumulative focus of terror beyond all terrors can be attributed to a single horrific entity: the Leviathan. Its title is a descriptor, with the root of the original Hebrew word possibly meaning "coiled" or "twisted."[2] Some biblical books, like Psalms and Job, reference Leviathan as a powerful and wild animal, subject only to its Creator. Other books, like Isaiah and Revelation, use the Leviathan as a symbol of hostile evil powers or the devil.[3] Biblical mentions of the Leviathan serve both to describe a literal creature and to create a memorable metaphor.

Leviathan may have looked like a creature from Julie Dillon's illustration *Beneath the Surface*.[4] A single boat sits calmly on flat waters—beneath which the viewer can see an abyss of crowded, ginormous, and utterly horrifying creatures. The widespread distribution of this original print demonstrates the substantial impact this piece of art has had on a variety of audiences. The image is simultaneously soothing and profoundly disturbing. It is my favorite monster illustration.

I've wondered about the little boat in Dillon's artwork. Is it aware or naive of its impending doom? Is the boat motionless in terror? Will it skirt the Leviathans undetected? The spectacular awfulness of these creatures always attracts my immediate attention, causing me to wonder why the surface is so calm. Does this signify breathless silence before the inevitable strike? If the Leviathans aren't attacking, is there a reason for that? Are they real or imagined? Is their ferocity real or imagined?

Indeed, is Leviathan truly an evil, heathen monster poised to slaughter any and all? Or is Leviathan rather a remarkably powerful and fearsome creature created uniquely and intentionally by God to be subject only to himself?

The Bible speaks as casually of Leviathan as it does of fish, birds, or flowers. That is to say, it references this ancient monstrosity as if it were no less real than any other created thing. However, Leviathan is clearly unique from the rest of creation. While incredibly powerful, called "the swiftly moving serpent, the coiling, writhing serpent" (Isaiah 27:1) and "the dragon" (Revelation 20:2[5]), Leviathan is also regularly referenced to illustrate God's strength, knowledge, and overall authority.[6] In other words, Leviathans are used to portray God's sovereignty.

> How many are your works, LORD!
> In wisdom you made them all;
> the earth is full of your creatures.
> There is the sea, vast and spacious,
> teeming with creatures beyond number—
> living things both large and small.
> There the ships go to and fro,
> and Leviathan, which you formed to frolic there.
>
> PSALM 104:24-26, NIV

We learn numerous things about Leviathan from these verses—and we also learn about God. Leviathan is a creature made by God, belonging to God. And apparently, Leviathan "frolics," a verb that connotes a free-spirited movement infused with joy, as if the creature praises and glorifies God simply by existing. The same psalm continues:

> All creatures look to you
> to give them their food at the proper time.
> When you give it to them,
> they gather it up;
> when you open your hand,
> they are satisfied with good things.

When you hide your face,
 they are terrified;
when you take away their breath,
 they die and return to the dust.
When you send your Spirit,
 they are created,
 and you renew the face of the ground.

VERSES 27-30, NIV

It becomes evident that Leviathan, a notoriously large, terrifyingly powerful creature that would strike fear into the hearts of even the bravest people, is completely at the mercy of God. Leviathan submits before God, looking to him for food and satisfaction. When God turns his face away, Leviathan—the bravest and most terrifying creature—becomes humble and terrified, knowing that in an instant God could reduce it to dust, as easily as he created it. The strength of this creature demonstrates by extension the much greater strength of the Creator.

You split the sea by your strength
 and smashed the heads of the sea monsters.
You crushed the heads of Leviathan
 and let the desert animals eat him.

PSALM 74:13-14

As powerful as Leviathan may be, its Creator God is far and away the most powerful of all. In the book of Job, Leviathan is also mentioned in reference to God's all-knowing, all-powerful nature. Job has suffered greatly and brings his complaint before the Lord. God responds with a series of rhetorical questions, a disciplinary rebuke for interrogating an omniscient, omnipotent Creator regarding things beyond Job's understanding.

> Can you catch Leviathan with a hook
>   or put a noose around its jaw?
> Can you tie it with a rope through the nose
>   or pierce its jaw with a spike?
> Will it beg you for mercy
>   or implore you for pity?
> Will it agree to work for you,
>   to be your slave for life?
> Can you make it a pet like a bird,
>   or give it to your little girls to play with? . . .
> If you lay a hand on it,
>   you will certainly remember the battle that follows.
>   You won't try that again!
> No, it is useless to try to capture it.
>   The hunter who attempts it will be knocked down.
> And since no one dares to disturb it,
>   who then can stand up to me?
> Who has given me anything that I need to pay back?
>   Everything under heaven is mine.
>
> JOB 41:1-5, 8-11

These verses proclaim the right order of things. If humans respect and stand in awe of the mighty power of Leviathan, how much more so they ought to stand before God! Job replies to this rebuke with complete humility: "I know that you can do anything, and no one can stop you" (Job 42:2). Although we may not have faced Leviathan ourselves, we can imagine attempting to subdue a sea monster only to regret it during the overpowering struggle that would follow. It is right for us to keep perspective of what creation teaches us about our Creator, and to respond to God as Job did. If we can acknowledge the greatness of a creature, we must expand our perspective of its Creator. Remembering Leviathan reminds us there is someone far greater than the fiercest being we will ever encounter.

I wasn't very far into my journey of becoming a marine biologist when I met my first shark. I was taking a coral reef ecology course in Belize, elated to be conducting snorkel fieldwork in the relative vicinity of the Great Blue Hole, which began as a giant sinkhole during the last Ice Age. At that time, a piece of land fell inward on itself (much like a cake might do in the oven), creating a depression roughly one thousand feet across. As time and weather eroded the hole further, unique geologic features formed out of limestone—similar to what you might find inside a cave. When sea levels rose, the land was eventually usurped by the ocean, and a coral reef grew around the perimeter. Today, the Great Blue Hole is defined by a shallow and colorful coral atoll surrounding the aquamarine waters of the hole itself, which sinks into a cave-like abyss that descends over 400 feet.[7]

I knew of the site from Jacques Cousteau's TV series *The Undersea World of Jacques Cousteau*. A famed marine biology icon and one of my inspiring role models, Cousteau named the Blue Hole one of the top five dive sites in the world.[8] When the surprise opportunity arose to dive there, I jumped at the chance.

I was a fairly naive diver at the time, new to the craft of perfecting my own buoyancy and unfamiliar with the shrouding layers of depth that gradually extinguish color. The sensations were more foreign than comfortable, and most of my time underwater, I held a vague if not wholly conscious sense of being scared. As such, my memory of the dive seems somewhat muddled, yet distinct for its novelty at the time and the unique characteristics of the experience.

From the boat, we floated on a distant sea of crystal blue. I could imagine the water darkening to an edge around us and I visualized in my mind's eye the circumference of an incredible circular mass lying just beneath. With a walk-in jump from the boat's side, we each splashed into the water, bobbing at the surface until all divers were assembled. I was the last diver to leap before we descended into blue.

I once heard it said that our thoughts of God are the size of a teacup of water, and God's thoughts of us are an ocean. Falling gradually deeper into the Blue Hole, I felt I must be inside God's cosmic teacup. It was so vast that I couldn't see from one edge of the cup to the other. Floating in bubbly stillness, all around me was otherworldly, and yet I felt myself the alien.

Some way down from the surface, the nearest wall of the Hole appeared at a distance, sloping up along our backs, sliding down into a blackened abyss. We followed the edge of the atoll, the round enclosed reef system, which appeared smooth at first sight but grew rugged as we neared. The forefront of my view was occasionally obstructed by meandering silver fish, flicking their tails through their backyard. With their eyes designed for the watery world, it's strange to realize that passing fish could see me more clearly than I could see them. To my eyes, they were shimmering shards of darting silver, yet they saw my black streamlined silhouette against the dazzling light from above. We silhouettes performed buddy checks, giving each other one-handed okay signals before diving deeper. The blue world swirled and slid past us.

An astonishing thing occurs the deeper you go: Colors disappear. Light does not function the same at thirty feet as it does at sixty.[9] It cannot penetrate through deeper water and retain the same amount of vibrancy, so the farther down you go, the more colors are lost. Reds, the longest wavelengths with the lowest energy, are the first to disappear. What should be a brilliantly peachy-red squirrelfish with large liquid eyes now looks a dusty brown. Then oranges and yellows vanish. You look at your hands and they are somehow whiter than a blank piece of printer paper. Somewhere along the way, purples and greens become blues. And then somewhere around 80 feet, you realize blue is the only color you have left. The world is formed by a spectrum of white to blue to black.

As colors faded on this dive, hints of nitrogen narcosis ebbed at the edges of my mind. Nitrogen narcosis is a physical phenomenon that can begin to affect a diver anywhere below seventy feet. The pressure of the overhead ocean water forces a greater amount of pressure on the human body and the breathing gas carried in the cylinder on their back, resulting in more nitrogen forced into the diver's body.[10] As this excess nitrogen circulates inside the diver, it begins to act as a kind of anesthetic in the brain, frequently heightening emotions while deadening physical faculties—an experience that divers call "being narced." Some divers have compared the feeling to that of being drunk, but the overall consensus is that narcosis "carves order from [a diver's] brain."[11] It is the evil ingredient in the recipe for disaster. It can numb a diver's awareness of something truly dangerous, causing them to take uncalculated risks. Conversely, it can also instill the irrevocable belief that something harmless will kill, resulting in panic. Nitrogen narcosis can be the deadly turning point. The only way to reverse its effect is to ascend calmly toward the surface. All deep-sea divers have experienced nitrogen narcosis at some point in their diving careers, and over time, the best divers learn to combat its effects through mental self-talk, extreme self-awareness, and even meditative techniques.

The Blue Hole was my first taste of nitrogen narcosis. I thought I was hearing whispering voices, overtopped by the banging of "jungle drums," my own pulse in my ears. My hands twitched with paranoia along my gear, checking and rechecking every strap, snap, and band. My brain became a blur of blue. Factoids I'd absorbed from a soggy brochure on the boat swam through my mind, the geomorphic reality now staring me straight in the face. "The Hole is roughly one thousand feet across and over four hundred feet deep. This feels like being inside an upside-down volcano! The wall before me is made of limestone, laced with icicle-looking stalactites that formed prior to the cave's full submersion, before sea level rose above the ancient reef. Those stalactites are big!" I watched another diver swim between them, as though the stalactites were massive Roman colonnades tipped in points. Following behind, I felt as though I was zigzagging between a massive set of carnivorous teeth.

Glancing to the side, I saw a distant silhouette. It was not a diver or anything else I had seen before. Sleek, fast, slicing through open water as if it was patrolling for intruders. The sharp pointed fins and whiplike tail . . . *It's a shark!* My nitrogen-intoxicated brain spewed every emotion from the old recurring, fear-fueled nightmare. My heart rate quickened, causing me to breathe faster, gushing a stream of rapid bubbles from my regulator. To calm myself, I began to hum an old, familiar song. My dive buddy waved at me: *Are you okay?* I nodded and gave the okay sign back, realizing I needed to focus. I was narced. I needed to be rational now. I grabbed at the most rational solution my narced brain could think of. I turned my back on the shark. If I couldn't see it, maybe it wasn't there.

We continued weaving our way among the stalactites before ending the dive. Slowly, making decompression stops along our ascent, I felt the narcosis lift from my brain as the world around us gained color again. And there was the shark, still off in the distance. Braver this time without the noise of jungle drums in my head or the chemical paranoia coursing through my veins, I looked at it.

It was a lemon shark, soft and solid against the blue backdrop. Several friends joined it, seeming to dance through the open space. I stared as they stealthily sped away from us along the reef's edge. Suddenly, I remembered what I had been singing a few moments before—lyrics that spoke about the wonders of the created world and how God is sovereign, creative, and kind. I'd sung them only to comfort myself in a fearful moment, but now within sight of the sharks' fading silhouettes, the meaning resonated with greater depth. When I had been afraid of creation, I was consoled to remember the truth of the Creator.

We dove again that afternoon at several shallower sites. At Half Moon, a torrent of blue-faced reef fish poured over us like a living waterfall. We spotted sponges the size of couches and bouquets of neon pipe corals, curlicuing out from the vertical substrate. A blacktip shark, about eight feet long, passed by, coming toward us for only a brief moment before being surprised by our presence and swarm of gurgling bubbles. The blacktip seemed more afraid of us than I was of it, disappearing as swiftly and silently as it had appeared. Truly majestic.

At Lighthouse Reef, I encountered a friendlier shark. A ten-foot-long bull shark coalesced in the misty distance, cruising directly toward our dive group. About twenty feet away, it slowed, taking us in with its dramatic white and black eyes. I felt as though I were seeing a tiger in its wild jungle home, pacing slowly before us. So close, so dangerous, so strong and beautiful. The shark's proximity was hair-raising, spine-tingling, and wonderfully gorgeous. I took in the distinctness of its gray back and white underbelly, the rigid sharpness of its dorsal fin, and the glimmer of recognition and curiosity in its eyes.

As the bull shark passed us and began to cruise off into the mist, I waved goodbye. It turned sharply, as though I had beckoned it to return. And it did come back to us, swimming with intensity and intentionality. Its body was enormous and thick with strength and speed. I shrank back toward the reef, hoping to tuck myself behind an outcropping of coral. Curious, the shark paced back and forth, eyeing our fins and our bubbles, even looking directly into our faces. Eventually, it moseyed off into the mist again. And when the fear dissipated, I felt as though I had been given a gift.

The most beautiful shark I have ever seen was off Pelorus Island in Queensland, Australia. I was operating a small boat for a team of volunteer divers from the Earthwatch Institute, an international environmental nonprofit. We were assisting with research on black band disease by collecting coral samples. The divers were swimming at the surface through the crystal water from our anchorage to the dive site. With one foot on the boat's edge and one hand gripping the awning overhead, I leaned out from the boat to watch them. A movement, just several inches from my foot at the edge of the boat, caught my eye.

An enormous, teardrop-shaped creature the length of the boat undulated side to side as it approached, its beautiful tan-brown skin mottled with dark speckles. Its dorsal fin slipped through the surface, setting off patterns of ripples. As a friendly dog approaches an unknown object, it was as if this creature sniffed the metallic edge of the boat, its body tilted as a large white and brown eye gazed upward at me. We locked eyes. Although I'd seen them only in taxonomy books and episodes of *National Geographic*, it was unmistakably a leopard shark.

I squealed with glee, which, of course, immediately frightened the friendly shark. It vanished with a gliding flick of its gorgeous spotted tail. My triumphant shriek had also alerted the divers. It was too late for them to swim back to see the shark themselves, but everyone craved to know the details, which I hollered from a distance. The divers began their descent hopeful of seeing the leopard underwater, while I watched their bubbles from atop the surface, thinking back on the astonishing beauty of the wild creature.

I have seen many more sharks since then, swimming with more than I can count or remember. But I have not had the recurring nightmare since my first shark dives in Belize. Real-life experiences have overwritten fear. Some sharks are curious, some are easily frightened, but all are remarkable creations, beautiful in their own right. They remind me that when we turn to face God, our greatest fears are quieted in his presence. Remembering who God is helps me to see the truth of how creation is designed. He has made all things "very good," including all the living creatures that teem in the water.[12]

God is sovereign. He is the Creator. He is far greater than what he has created. While our world today may not contain the living

Leviathans of ancient times—the sea dragon ichthyosaur or the haunting sea monster renderings in *Beneath the Surface*—it does have sharks, whales, and giant squid. Praise God for such beautiful works of strength and stealth and speed, created with intentionality and joy! Their existence teaches us much of their Maker.

5

living water

Where silty sand stretched as far as the underwater eye could see, Davis Rock in Dry Tortugas National Park rose thirty feet toward the surface, as if it were a single massive conglomerate rock half-buried in the seafloor. The Rock itself was the length and width of approximately two city blocks. Divers could survey the perimeter of Davis in thirty minutes, so long as the current was not too strong. When the current flowed, water ripped along the windward side of the Rock, threatening to shove tired divers out to open sea. But in good weather, the Rock was a treasure trove of living motion and color.

Davis Rock was coated in corals. Massive green and yellow *Orbicella faveolata* colonies rose in six-foot-tall mounds across its surface, mysteriously reminiscent of Stonehenge. These lent a sacred sort of feeling, inspiring quiet awe as I meandered between them. Sometimes I'd notice the giant bubble-polyped *Montastraea cavernosa* corals fluorescing blue, orange, and pink. Wedged between crevasses, bouquet-like clusters of yellow *Eusmilia fastigiata* glistened in mottled light filtering down from the surface.

Flashes of life skirted past my field of view. Every dive at Davis Rock was a fishy one, but some dives introduced more fish than others. Jubilant schools of amberjacks danced above coral colonies in blurs of yellow and silver while extended families of blue tangs swirled over rocks, picking off bits of green algae. Gangs of speckled, fang-toothed barracuda glistened white as they patrolled the perimeter. Vertically striped sergeant majors filed in attention past my vantage beside a bed of rippling macroalgae. The place was a flurry of motion with streams of fish circling, rising, dodging, lingering, and chasing their way through the neighborhood.

Composed of far more than fish and corals, this community was bursting with inhabitants. The sea anemone thicket jiggled in the current, gelled tentacles tangling and unsticking, grasping at plankton particulates and working food toward their flowerlike mouths. A dark green, thick-bodied moray poked its insane-looking toothy smile and unblinking white eyes from its slithery hideaway home in the Rock's underside. Fuzzy globs of brown-green algae bounced loose from their growths to scatter and settle across the reef. Miniscule striped shrimp with threadlike whiskers scuttled from minute clefts to graze and clean and garden. The sleek silhouette of a prowling reef shark jettisoned itself toward the Rock before disappearing into the silky mist.

Every time I came to the site, there was something new to see. Some new genre of lively magic in this underwater world. Some new character of the marine neighborhood elected to greet me in its own particular fashion. Uniqueness galore! The place was always the same, but its activity was always different. Davis Rock was a glistening, moving, active, and very alive place to visit. A healthy reef is predictable only in its diverse unpredictability—a place that feasts on the riches of resources offered by the corals that build it. Bedrock of ancient coral skeleton, skinned over with live and vibrant tissue, enjoyed by diverse multitudes of creatures. Everything at Davis was wild and free, overwhelming the senses in an ecstasy of splendor. Facets of this integrated, uniquely combined community were only faintly revealed amid its dynamic, flowing, and unceasing nature. I could dive the site a million times over and never see the same scene twice.

A healthy reef is remarkable in its aliveness and is similar to what we might expect to find from a certain book, likely sitting on some lofty shelf in our homes, hopefully speckled with more fingerprints than dust. This book is actually a compilation of sixty-six books written by forty authors from three continents over a two-thousand-year period.[1] It is a trustworthy source because it's inspired, infallible, and inerrant,[2] going by several different names: the Bible, the Word of God, and Scripture. The Bible can be compared to a healthy reef such as Davis Rock for several reasons.

There was always something new to see at Davis, and the same is true when we read the Bible. It is living and active, like schools of fish swirling over a reef, studded with colorful corals and swathed with diverse algae moving in the current. Come to it once, and you will see the family of blue tangs nibbling at seagrass. Come to it a second time, and you will catch sight of a thicket of iridescent anemones unfolding their photosynthetic tentacles toward sunlight. Come yet again—there will be a shark or ray gliding over sand beyond the Rock's edge. There will always be something to discover and treasure, inspiring wonder and learning. Out of the water, you can almost forget the reef's beauty. But as you visit more and more, you will find yourself slowly falling in love with Davis Rock, even daydreaming about it from land. Just as Davis Rock is bursting with life and movement, "the word of God is alive and active" (Hebrews 4:12, NIV).

Of course, the more frequently a diver visits a reef, the more familiar they become with it. Certainly, there may always be something new to see, some standout event that occurs once within their viewing, some extraordinary reason to remember that particular dive as special. The reef may continue to surprise them as they discover new facets of its character, personality, and story. But through exposure, they garner a deeper level of knowing. When a diver spends a lot of time on a reef, they learn where the blue tangs swarm over algal beds, which cleft of rock houses the crafty green moray eel, and that a shark sighting is more likely on the north side. The tangs might be moving in a different arrangement from one place to the next, the moray might be surprisingly shy one day compared to another, and divers might be startled to see two sharks instead of one, but the fact is that they know the reef. When a diver has witnessed its treasures before, they come expectant to discover them yet again, in all their comforting familiarity and wonderful freshness.

In the same way that a diver can expect to see amazing things at Davis Rock, we can expect to find treasures within the pages of the Bible. But do we approach it this way? Do we come to its teaching with the diver's sense of expectant discovery? Are our eyes even open when we delve into it? Or are we swimming through it distracted and apathetic, unable to see the beauty, meaning, and life all around us?

The Rock's reefscape shimmers with color and beauty, requiring care and attention to remain vibrant. Likewise, the concepts of the Bible swirl around our lives like fish over a reef, moving and dancing, chipping away at little bits and pieces, pruning back invasive weeds, and relishing in the sunlight. Scripture, we're told, is "sharper than any double-edged sword, it penetrates even to dividing soul and spirit, joints and marrow; it judges the thoughts and attitudes of the heart" (Hebrews 4:12, NIV).

The Bible explains its own importance and usefulness. "All Scripture is God-breathed and is useful for teaching, rebuking, correcting and training in righteousness, so that the servant of God may be thoroughly equipped for every good work" (2 Timothy 3:16-17, NIV). The Bible is breathed by God—an interesting image to consider both literally and figuratively. The words the Bible contains teach, rebuke, correct, and train with the purpose of equipping readers for "every good work." When we approach the Bible with the mindset that it has power to equip us in these ways, we engage with the text differently. What may initially seem like an ancient manuscript translated from Hebrew, Greek, and a bit of Aramaic (which it is) becomes so much more; it becomes applicable to our lives. It becomes engaging, even conversational. The Bible has the ability to speak into our lives with the power of God to change us. We ought to consider whether we approach it with this expectation. But there's more.

Through continued exposure, Scripture can become like a favorite dive site. As much as we may come to recognize its words, it unceasingly offers up surprises from its pages—treasures of discipleship to teach, rebuke, correct, and train us so that we might become equipped for every good work. Reading the Bible blesses us. And when we open its pages, we have the opportunity to marvel at the beautiful perspectives of creation and the Creator that it offers. In fact, as you read the Bible more and more, you may find yourself slowly falling in love with the one of whom it is about—Jesus.

Just as the Bible is also called Scripture and the Word of God, Jesus is referred to in different ways. He himself was a big fan of speaking in parables and figurative language as a way of expressing complex cosmic concepts in small, understandable bites. One such example connects back to our theme of Davis Rock's active and life-filled waters. That's because water was one of Jesus' favorite metaphors to articulate the impact and meaning of his saving grace.

Jesus spoke often of "living water." He said, "Anyone who is thirsty may come to me! Anyone who believes in me may come and drink! For the Scriptures declare, 'Rivers of living water will flow from his heart'" (John 7:37-38). Another time while standing at a well, Jesus said, "Anyone who drinks this water will soon become thirsty again. But those who drink the water I give will never be thirsty again. It becomes a fresh, bubbling spring within them, giving them eternal life" (John 4:13-14). Unlike the Fountain of Youth mythology that has existed for millennia—popularized by the ancient Greek historian Herodotus and Spanish explorer Ponce de Leon and more recently in movies like *Tuck Everlasting* and *Pirates of the Caribbean: On Stranger Tides*—Jesus speaks of the way that water represents his influence in our lives. He is the source of living water, where people can find what actually satisfies. He will not make us thirsty; instead, he satiates our thirst and gives us something that lasts forever. If we require regular water to survive, we need the superior "water" that Jesus offers even more! What would it look like if, instead of turning to whatever is metaphorically represented by normal water, we received that which truly satisfies from Jesus?

The words of Jesus aren't the only times we see the concept of water used to describe the relationship between our thirsting need for what satisfies and Jesus himself. The book of Revelation describes an incredible otherworldly picture of the Kingdom of God in this way, speaking of "a river with the water of life, clear as crystal, flowing from the throne of God and of the Lamb [Jesus]. It flowed down the center of the main street" (Revelation 22:1-2). These verses imply that on some level, the living water offered by Jesus isn't purely metaphorical. In his Kingdom, there may be a literal river, clear like crystal, flowing from his throne.

Overall, these strong biblical references to living water teach us that Jesus himself is the source of the real water we need. While we may think we thirst for one thing, we truly long for what only he can provide. Only he can truly satisfy. And his version of living water is truly alive. The Word of God, Jesus, also provides us with the Word of God, the Bible, which is as living and active as the reefscape of Davis Rock.

The Bible can teach us a lot about the character of God by describing his role as Creator. It is important to know that "long ago by God's word the heavens came into being and the earth was formed out of water and by water" (2 Peter 3:5, NIV).

For a moment, let's focus on "God's word." The Gospel of John begins: "In the beginning the Word already existed. The Word was with God, and the Word was God. He existed in the beginning with God" (John 1:1-2). "The Word" here does not refer to the Bible (which, in the beginning, had yet to be written), but rather the literal word of God that enacted his creative power.[3] This reference to the Word is special. Not only does it refer to the literal speaking God did at the beginning of creation, but it also serves as a figurative way of saying that Jesus, the Son of God, was present with God the Father in the very beginning. We can know this from the verses that follow John 1:1-2, which go on to describe Jesus and reveal an interesting connection between the Bible as the Word of God, being breathed by God, and the Trinity.

The concept of the Trinity is considered core doctrine to the Christian faith. In other words, it is a foundational teaching that informs much of our understanding of the Bible and the character and nature of God as the three-in-one entity of God the Father, Jesus the Son, and the Holy Spirit. The doctrine of the Trinity teaches that each of these persons is holy and wholly unified despite having different divine roles. When Christians use the name God, this triune God is the one of whom they speak.

Acknowledging that our God is simultaneously one singular God composed of three persons shapes our understanding of who God is and how the Father, Son, and Holy Spirit act and move together in unity. While the doctrine of the Trinity is frequently discussed by biblical scholars, I would like to focus on something I refer to as the "ecology of the Trinity"—what we can know about the relationships within and between the Trinity and everything else.

In other words, what does God's triune nature reveal about his creation? We can look at how the Father, Son, and Holy Spirit are described and consider how this relates to the created world. While we have already discussed the roles of Father and Son, we have not yet considered the role of the Holy Spirit. To do this, let's revisit some passages.

As Jesus was present with God the Father in the beginning of creation, so was the Holy Spirit. We know this because the Bible names the Holy Spirit as being there. When "the earth was formless and empty, and darkness covered the deep waters . . . the Spirit of God was hovering over the surface of the waters" (Genesis 1:2).

Additionally, there is a curious sort of conversation that occurs in Genesis 1, the poetic summary of the creation account. God says, "Let us make human beings in our image, to be like us" (Genesis 1:26), which contains an interesting reference to a self-described "us." It's as if multiple persons are present—because they are. What follows is that "the Lord God formed the man from the dust of the ground. He breathed the breath of life into the man's nostrils, and the man became a living person" (Genesis 2:7). The Hebrew word for "breath" in this context is the same word that was used to describe "Spirit." We might ponder this account with the imagery of the Holy Spirit breathing himself into the man, setting him apart from other living creatures by being made in God's image.

Once we understand the role of the Holy Spirit in creation, we can consider how he works through Scripture. Jesus, the Word, is revealed in and through the Bible, and amazingly it is the Holy Spirit who actually does the revealing. If Jesus is the subject, the Holy Spirit is the verb. "All Scripture is God-breathed." The Bible "is useful for teaching, rebuking, correcting, and training in righteousness" (2 Timothy 3:16, NIV) to equip its readers for every good work because the Bible itself is infused with the Holy Spirit. The reason the Bible is alive and active is because the Holy Spirit interacts with us as we read the text. He also is the one who enables us to see and engage with God through his creation.

I hardly remember my first time diving at Davis Rock. I just recall that the weather was poor, the current was strong, and the visibility was low. Kicking hard to stay in place, I could barely see more than ten feet in front of me, which felt a bit eerie. My dive buddy was a fuzzy shape in the milky, particle-filled water at my side as we descended on a site we couldn't see. And then suddenly out of the haze, a dark, ground-like mass appeared beneath us. Drawing nearer, the textures and colors of Davis Rock came alive beneath my gaze. Perhaps the rest of that particular dive day is forgettable because I saw hardly anything of note. But when I dove there again in the days that followed, I came to know Davis Rock for the special beauty it is.

It is a curious thing. Coming to know God is a lot like that. It is much like encountering a place or meeting a person for the first time. When you meet someone, you might not have much to say right away. Conversation might feel a little hard, even uncomfortable. You converse on shallow topics and might not stay in the person's presence very long. But just as it takes time and effort to get to know a person, it takes time to really know God. You listen to what God says as you read his Word. You're honest about yourself and your own life, gradually going deeper as relationship develops. You visit with each other more often, making a plan to intentionally meet up again later. You approach your time together with a genuine sense of expectant discovery. After a while, it becomes hard to imagine not having him in your life anymore, impossible to see your future without him. You long to be together again. You are best friends. And then, as you continue getting to know God more and more, reading the Bible, spending prayer time in his presence, and savoring his creation and what it teaches you about his character, you start to really love him.

Relationship with God is deeply personal. No two relationships are identical, and so it should not be surprising that each person's approach to knowing God may look a bit different. I believe God makes himself known to each of us in the way we will best understand. For me, this takes the shape of spending a lot of time outside, praising God in the cathedral of his creation and learning more of his character by studying what he has made. I delight to identify his fingerprints in places like Davis Rock, especially after so many years of diving and reading the Word. But what about the person who feels uncertain, who wants a relationship with God but has zero idea of how to start?

This is what I recommend: Soak in the world God has made. Marvel at his creation. Consider what it can teach you about him. Learn the limitations of where it can no longer instruct, where you must take your questions back to his Word and directly to him in prayer. Ask the hard questions. Let your own understanding be challenged and turn to him for the Truth. Know that there will always be something new to discover. But most importantly, drink in the Word of God, as though it is living water. Where you are thirsty, turn to Jesus. Make a habit of staying spiritually hydrated, reading the Bible regularly. Remember that relationships start with an introduction and deepen with time spent together. As you invest this time and your relationship grows, you will come to sense the Holy Spirit "breathing" on you as you gain greater understanding of who God is and what the Bible teaches about him. Immerse yourself in his world and his Word, the source of living water.

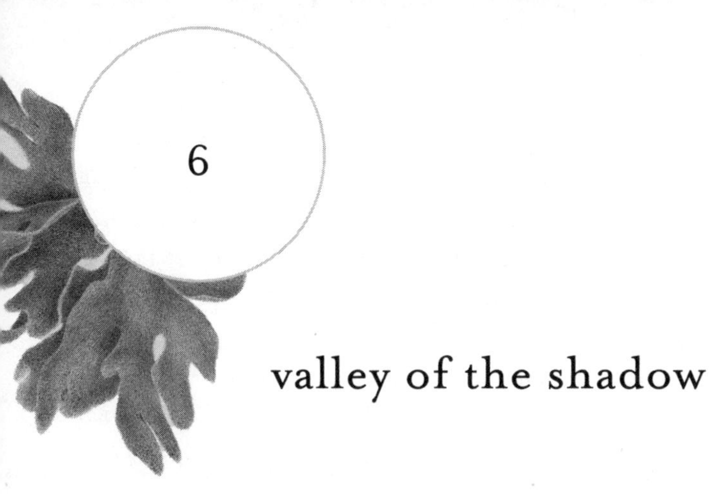

6

valley of the shadow

The coral was older than my great-grandmother and the size of a Volkswagen Beetle. Such gloriously large colonies are becoming increasingly rare along the Florida Reef, so from the get-go I thought of this *Meandrina jacksoni* as special. The little fish swirling in graceful dances around her seemed to think so too. Her colorful ridges were the shade of love at first sight.

Coming back to this colony month after month was like visiting a wise, well-respected friend. Floating in the water beside her, I watched the light move in dazzling lines across her wrinkled face. Seeing the way crabs, Christmas tree worms, and minnows frolicked to her, I couldn't help but feel she had some bit of sage advice to offer me. If corals could speak, undoubtedly she would have told tales of storms that raked the reef, shaking the house that she (as so many before her) strove to

build. She would have reported on the frisky little stingray who caused daily mischievousness over the nearest ridge. She would have kindly asked why I swam so hard. Was I trying to outrun sharks? Was I ravenous for food? And after I answered no to both questions, she would graciously tell me to slow down. To take in the light, colors, and textures offered to me right here. To please spend a quiet, ceasing moment by her side.

And so I did.

I visited the grandmother coral regularly for more than a year before she contracted SCTLD. She required hospice care soon after. Despite my doing everything I could, the disease took her quickly. She fell silent in its grasp, and within four months, she was gone. One day she was grandmother to the reef, and the next she was stone. I miss our moments of conversation and stillness together.

Everything that has ever lived has walked this path before us, and everything to ever live is promised the same future. "Death is the destiny of everyone; the living should take this to heart" (Ecclesiastes 7:2, NIV). Since we take such great care to learn how to live well, we should also consider what can be learned from life's ending. Perhaps it will make us wiser when our day to return to dust comes. In order to "take this to heart," we need to understand the meaning of death, both from biological and spiritual perspectives.

Some people may be skeptical of the biological sciences as a means for understanding death; similarly, some may be skeptical of looking to the Bible to comprehend death. That is very well and good. We ought to be critical in our examination of what we believe and why. And we should learn to better hear and understand reasons for why others may think differently.

I must note that for those left behind, grief can be one of the most acute and long-lasting hardships to endure. Grief is the agony of love unable to reach its destination. It has been said, "Grief changes us. The pain sculpts us into someone who understands more deeply, hurts more often, appreciates more quickly, cries more easily, hopes more desperately, and loves more openly."[1] As the band Switchfoot wrote in their song "What It Costs," "Love deals the currency of loss."[2]

All grieve as surely as all die, if not now, then presently. It is not my intention to undermine or ignore either the validity or intensity of grief; I, too, have tasted its bitter dregs and judge nobody in such a position. Condolence is not enough, so I will not attempt it here. But we might compare the sharp stinging and benumbed aching to symptoms of growing pains. We who grieve are like confused children crying out in the night because of internal aches in our legs, as muscles build and bones calcify layer upon layer. We awake in the morning to find that, for all our pain, we are one millimeter taller. In several years' time we discover that we have grown many inches. Far more unspeakable pain marks the path of grief, but having acknowledged this, let's look first at the place of death in evolutionary history and its continuously unfolding usefulness in ecology.

Science is frequently accused of being sterilized of feeling, as it rightly ought to be. For if science were emotional or came with preconceived ideals, it could not be trusted. "Ignorance is the natural state of mind for a research scientist. People who believe they are ignorant of nothing have neither looked for, nor stumbled upon, the boundary between what is known and unknown in the universe."[3] The neutrality of science's approach to topics such as death lends to its credibility.

Death is one kind of change. Changes occur over time. We see this pattern in every aspect of the created world: the growth of a forest, the mounting power of a volcano, ever-moving cloudscapes, the expansion of galaxies. Some changes seem big, looking bigger when we view them from the vantage of large time frames and intentional hindsight. But most changes begin in tiny ways, such as the mutation of a single nucleotide within a genome (the accidental changing of a single letter in the several-billion-long letter strands of DNA). Most of the time, tiny changes are neutral, having no visible result.

Biologically speaking, as in the instance of a single changed nucleotide, changes may be beneficial or detrimental to an organism's life. If the change is problematic for an organism's survival and reproduction, the change probably won't be passed to the next generation; as quickly as the change appears, it dies with the organism. Changes that are neutral or helpful to an organism's survival and reproduction are passed to the next generation. When beneficial changes are handed down, generation after generation, it can help a population of organisms become more successful in their environment. This process of tiny changes that either hurt or help an organism's ability to survive and reproduce in its environment is called natural selection. When miniscule changes slowly accumulate over extremely long periods of time, the outcomes can look and function differently from their predecessors. Even Charles Darwin observed these changes with wonder: "There is grandeur in this view of life, with its several powers, having been originally breathed into a few forms or into one; and that, whilst this planet has gone cycling on according to the fixed law of gravity, from so simple a beginning endless forms most beautiful and most wonderful have been, and are being, evolved."[4] Change happens over time. This is the theory of evolution.

It is critical to note that evolution is never forward-thinking. It is not philosophical. It is no religion; it cannot be moral, and it does not plan. It is simply a principle of pattern.[5] In the same way that gravity acts on all things, regardless of what that thing is, evolution cannot pick and choose. It has no end destination in mind. The changes it enacts are purely random and altogether arbitrary. It is only when we, as scientists, observe a change's outcome within a given environment as helpful, hurtful, or neutral that we recognize which organisms are advancing toward survival or death. As such, evolution cannot in any way substitute for the role of Creator. At best, it serves as a tool by which the Creator may choose to create. (Most certainly, thoughtless evolution has no end destination in mind, but the Christian knows that our thoughtful Creator God does.)

The theory of evolution is supported by an ever-expanding body of studies, including evolutionary biology, comparative genomics, paleontology, and anthropology, to name a few. There is a reason why it is the current scientific consensus. Anyone who refutes such highly supported ground should do their homework carefully before stepping into an argumentative arena of their own making. As noted by theologian John H. Walton, "Though we should not blindly accept the scientific consensus if its results are questionable on scientific principles, we can reach an understanding that regardless of whether the scientific conclusions stand the test of time or not, they pose no threat to biblical belief."[6] The theory of evolution is indeed based on well-founded scientific principles. Regardless of how we view evolution in years to come, well-researched science and well-understood bibliology are in no way mutually exclusive. In fact, one quite often points in the direction of the other—and both point to the Creator.

Evolution (change over time) plays a crucial role in ecology, which, as previously mentioned, is the study of relationships between biological organisms and their environments. As a branch of biology, ecology endeavors to understand how and why living things function in certain places. All living things play specific and oftentimes special roles in their environment, called niches. Ecologists are obsessed with niches. While a biologist may be an expert on trees, an ecologist also wants to take a good look at the whole forest.

Biologists view evolution through the lens of how it leads organisms to change over time, and when biologists see death, they primarily note the cessation of life and the beginning of decomposition. Ecologists see evolution and wonder how the cumulative changes acting on those organisms have in turn transformed whole ecosystems. When ecologists see death, they consider the niche of that organism (the role that particular life played in its environment) and wonder what will fill that niche (perform that role) now that it has been vacated.

When a niche opens (whether due to death, sickness, migration, etc.), there are consequences, much like when a house is put up for sale. In competitive environments, another organism will rush in to fill that niche. House sold. This is what typically occurs when a coral dies in a thriving reef ecosystem; algae and other corals in need of space and light rush to take its place. Less frequently, it takes time for another organism to fill an open niche, or that niche may remain open. Like an empty house slowly going to ruin, a vacant niche can cause all sorts of problems for an ecosystem if it's an important niche to have occupied. This is what happens when a species that contributes something special to its environment goes extinct or extant. For example, sea turtles are one of the few animals that eat jellyfish, gobbling them up like kids eat cotton candy. Unfortunately, nearly all species of sea turtles are listed as endangered. If we lose sea turtles, there is likely to be a dramatic increase in global jellyfish populations, unless some other jelly-eating animals were to rise to the occasion.

There are many documented examples of niche-filling processes, but one of the most well-known is the story of wolves in Yellowstone National Park. Gray wolves were hunted to the brink of extinction in the 1900s, being misconceived as a pest species because they beat hunters to their quarry. In his famous essay "Thinking like a Mountain," environmental writer Aldo Leopold explained, "I thought that because fewer wolves meant more deer, that no wolves would mean hunters' paradise. . . . I now suspect that just as a deer herd lives in mortal fear of its wolves, so does a mountain live in mortal fear of its deer."[7] The last wolves in Yellowstone were killed in 1926. As the field of ecology grew in leaps and bounds throughout the mid-1900s, the Endangered Species Act was passed in 1973. People began to question the extermination of wolves and consider, even amidst widespread controversy, the possibility of reintroduction.

Wolves were released back into Yellowstone in 1995, and their presence created a myriad of ecological changes.[8] Local elk populations began to browse more widely (from fear of predation), which in turn allowed willow trees (eaten by elk) to grow in places where they previously had not. This led to an increase in beavers (who use willow to build dams), which in turn altered the park's water systems (because of an increase in beaver dams) and created marshy habitats (new niches) for other species, including moose, otters, mink, waterfowl, amphibians, and fish. The presence of wolves seems to have restored the ecology of Yellowstone National Park to what it had been prior to their extermination, although there is debate as to whether the wolves' reintroduction was the only factor in its renewal.[9] Imagine this story in reverse, were the wolves removed from Yellowstone once again.

In my opinion, evolution is probably not the entire scientific story because science itself is always evolving. Our knowledge of the world is always expanding. There is much we have left to learn and various knowledge gaps to fill. We are likely missing a key component in our current scientific understanding—one that could further revolutionize or enhance the clear connection between the fields of science and the Genesis creation account. But even now, through the lenses of natural selection and ecology, we can see how death functions as a tool. In natural selection, death is the selector power of whether or not changes within an individual or population will be passed on. From an ecological perspective, death opens and mediates the availability of niches, thereby shaping the sorts of lives and interactions we see within an environment.

Strangely, corals were created with an ability to put off death. Like many of their cousins in the taxonomic phylum Cnidaria, corals do not undergo senescence, the normal aging process by which cells' replicating ability wears out toward the end of an animal's life, resulting in what is often referred to as a "natural death."[10] Unlike human cells, which naturally degenerate over time, the cells of corals continue replicating superbly, allowing them to live virtually forever.[11] Because corals secrete calcium carbonate to build their skeletons, biologists can tell how old a coral is by coring it and examining seasonal growth rings, much as arborists do with trees. The Great Barrier Reef is home to one of the oldest living coral colonies in the world, Muga dhambi, which is over four hundred years old.[12] While corals may not be age-limited like other animals, they do die. Disease, pollution, and rapid environmental changes (such as temperature, pH, and salinity) that create stress result in their deaths. It strikes as particularly tragic that an organism designed with unique death-defying mechanisms now experiences death regularly because of the world humans have altered around it. The grandmother coral I knew is one such example.

While biology describes death as a tool that shapes the world, the Bible describes its impact.[13] Death is the result of sin,[14] the choosing of something over God (Deuteronomy 9:7; Joshua 1:18; Romans 1:28; James 4:17; 1 John 3:4) or "to miss the mark."[15]

The account of sin entering the created world is both strange and disturbing. Sin entered through Adam,[16] the first created man noted in the Bible, when he chose of his own free will to eat fruit from the tree of which God had told him not to eat: "If you eat its fruit, you are sure to die."[17] Starting with Adam and emerging throughout the whole biblical narrative, we see the consequences of sin. These ramifications extend and apply to our lives today. As if irreversibly woven into the DNA of humankind, sin acts like a permanently hereditary allele or highly contagious disease. "Sin undoes the creation of beauty, order, and peace," says author and speaker Phylicia Masonheimer. "It is ugliness, disorder, and chaos."[18] The world changed from the good design God originally intended to a corrupted and unholy place, forever separated from its holy Creator—or so it would seem.

Enter Jesus. He is called the Son of God, being fully God and yet fully human. In an imperfect, sinful world, Jesus lived a perfect, sinless life and died a horrific death, crucified as an atoning sacrifice in the place of humankind (all of us sinners) so that by acknowledging him, we could be reconciled to God.[19] When we turn away from the things that separate us from God, and turn instead toward Jesus, the author and perfector of our faith,[20] we are promised that "he is faithful and just to forgive us our sins and to cleanse us from all wickedness" (1 John 1:9). When we choose to live through Jesus, we receive the Holy Spirit, the active presence of God moving and working in our hearts,[21] and the promise of eternal life, to dwell in the presence of God forever.[22] This is why the story of Jesus' life, death, and resurrection are described as the gospel, meaning "good news." If all of humankind has the opportunity to be fully reconciled to God and experience redeemed life on earth and restored life after death through Jesus, this is the greatest news of all!

But there's more. When we invite Jesus into our lives, we become children of God.[23] "Since we are his children, we are his heirs. In fact, together with Christ we are heirs of God's glory. But if we are to share his glory, we must also share his suffering" (Romans 8:17). Jesus took on our sin and suffered in our place so that we might have a share in his glory, in his relationship with the Father. And this life transformation of becoming God's adopted heirs connects back to the rest of creation.

> The creation waits in eager expectation for the children of God to be revealed. For the creation was subjected to frustration, not by its own choice, but by the will of the one who subjected it, in hope that the creation itself will be liberated from its bondage to decay and brought into the freedom and glory of the children of God.
>
> ROMANS 8:19-21, NIV

Creation waits in eager expectation for two things: for the children of God to be revealed and to be liberated from bondage so that it can be brought into freedom. By us sharing in the suffering and glory of Jesus, all of creation will be liberated from the power of sin and death. This means that the gospel is not only good news for humans created in the image of God; it is good news for all of creation!

There is more evidence for this when comparing Adam and Jesus. "The sin of this one man, Adam, caused death to rule over many. But even greater is God's wonderful grace and his gift of righteousness, for all who receive it will live in triumph over sin and death through this one man, Jesus Christ" (Romans 5:17). Adam's sin opened the door for death, but Jesus' sacrifice has basically ripped that door off its hinges and left the room beyond it in shambles. His salvation gives life to anyone who receives him. Again, we see the antithesis comparison of Adam and Jesus. "Just as death came into the world through a man [Adam], now the resurrection from the dead has begun through another man [Jesus]. Just as everyone dies because we all belong to Adam, everyone who belongs to Christ will be given new life" (1 Corinthians 15:21-22). Adam brought death, but Jesus brought life. As all of creation was included in the Genesis curse through the sinful act of Adam, so all of creation is included in the restoration made possible through the saving act of Jesus. If all creation was condemned in the curse, all creation will be restored in the blessing. This is profoundly exciting because if all creation is being redeemed and restored, then we can expect all creation to be involved in the eternal unfolding of God's story.

Someday, death will be defeated.[24] It may rule today, but Jesus has usurped its authority and destroyed its lasting power over us. Through him, all of creation can experience resurrection.

> It will happen in a moment, in the blink of an eye, when the last trumpet is blown. For when the trumpet sounds, those who have died will be raised to live forever. And we who are living will also be transformed. For our dying bodies must be transformed into bodies that will never die; our mortal bodies must be transformed into immortal bodies.
>
> Then, when our dying bodies have been transformed into bodies that will never die, this Scripture will be fulfilled:
>
> "Death is swallowed up in victory.
> O death, where is your victory?
> O death, where is your sting?"
>
> 1 CORINTHIANS 15:52-55
> (QUOTING ISAIAH 25:8 AND HOSEA 13:14)

It is from this firm foundation that I challenge the near universally held belief that death is "bad." Death has never been a problem scientifically speaking; it has been a necessary tool. Death is no longer in power from a biblical standpoint; Jesus conquered it, and someday "there will be no more death or mourning or crying or pain" (Revelation 21:4, NIV). For those who invite and receive the salvation of Jesus into their lives, the purposes of death and life are completely rewritten. Death is but a flimsy barrier between our current existence and the glorious future we're promised of forever existing in the intimate presence of our beautiful Savior. Death is no longer a mere functional evolutionary tool, but now also a portal to the embodiment and fulfillment of complete restoration. Life is no longer a meager window of opportunity for the self-seeking and self-serving behaviors of natural selection but instead has been transformed into a tangible opportunity for eternal relationship between created and Creator, out of which created life yearns toward entire sanctification. In life, we no longer fear death for ourselves, although we continue to feel the deep aches and laborious longings of a world under the influence of sin, where death separates us as we cry out for the future unity and restoration that only God brings.

Having had a front-row seat to the death of an ecosystem, having now seen the lives of thousands of corals suffer and end, I make the claim that corals are remarkably skilled at praising their Creator even as they die. There is no quantifiable scientific data to back this statement, for it is a spiritual claim. I have seen it with my own eyes countless times. For as long as it lives, a beautiful animal ravaged by disease continues to reach, sing, and outpour its created life into the hands of the Creator.

This is what we ought to learn: That to grieve well and to die well elicit free rejoicing and triumphant praises to God. That even in death, "praise is the mode of love which always has some element of joy in it. Praise in due order; of Him as the giver, of her as the gift," as C. S. Lewis noted. "Don't we in praise somehow enjoy what we praise, however far we are from it? I must do more of this. . . . And then, of her, and of every created thing I praise, I should say, 'In some way, in its unique way, like Him who made it.'"[25]

Sometimes I remember the grandmother coral's transformation from life to stone, and I can reverse the story in my mind's eye. While looking at the remnants of a dead reef, I can imagine how things used to be. I follow the structure of the calcium carbonate bedrock with my eyes, feeling its textures and ridges with my calloused fingers, and sense where life once was. I recognize patterns created by specific species. A cemented squiggle in the edge of a rock becomes the home of the brilliant orange feather tube worm that used to live inside a mustard-colored *Porites astreoides* colony. The gray sticks poking up from the sand turn back into the leathery, purple sea fans they once were, like lace vibrating in the current, swaying musically back and forth. The empty water column above the vacant reef becomes a whirl of color, as the ghosts of parrotfish and amberjacks, barracuda and damsels, angels and butterflies flutter among their homes.

The gift of being able to see things as they once were, when they were alive and good, can feel like a curse when you find yourself surrounded by stony skeletons and silence. I remember the prophet Ezekiel, who found himself brought to a valley full of bones, being led back and forth among them.[26] God asked Ezekiel, "Can these bones live?" (Ezekiel 37:3, NIV). As I swim between browned, algae-encrusted skeletons, over the piles of sand, fractured remnants from some reef of the past I will never see, I ask the same question. *Can these bones live?* I remember Ezekiel's reply: "Sovereign LORD, you alone know" (verse 3, NIV). Yes, God alone knows. I feel as Ezekiel felt, surrounded by death and grappling with confused grief, detached from and yet totally invested in the world where I find myself. I am the only living thing in this valley of the shadow of death, and yet I converse with the Spirit of God.

While swimming among the coral rubble, I remember what God told Ezekiel. "Prophesy to these bones and say to them, 'Dry bones, hear the word of the LORD!'" (verse 4, NIV). I see glimmers of fish that exist no more, colors of corals that lived long before I did. I remember the miracle in the valley of dry bones, when God spoke through a lonely human and brought the bones from death to flesh, from flesh to life, all to declare:

> I am going to open your graves and bring you up from them. . . . Then you, my people, will know that I am the LORD, when I open your graves and bring you up from them. I will put my Spirit in you and you will live, and I will settle you in your own land. Then you will know that I the LORD have spoken, and I have done it.
>
> EZEKIEL 37:12-14, NIV

In the valley of the shadow of death, I see not only remnants of the beauty that once existed. I see flickers of light, glinting off scales and colorful textures of the "swarms of living creatures . . . large numbers of fish" (Ezekiel 47:9, NIV) living in "the river of the water of life, as clear as crystal, flowing from the throne of God and of the Lamb down the middle of the great street of the city" (Revelation 22:1-2, NIV). No curse. All things new. Everything right.

Perhaps in that river connecting to the sea where the water turns fresh, I will swim in the presence of the Lord. Maybe as I praise him, he will lead me to the newly made, perfectly restored friend I once knew in the valley of the shadow. Maybe there, dazzling with light and glorious color, praising her loving Creator with every cell of her remade being, I will rest beside the grandmother coral again.

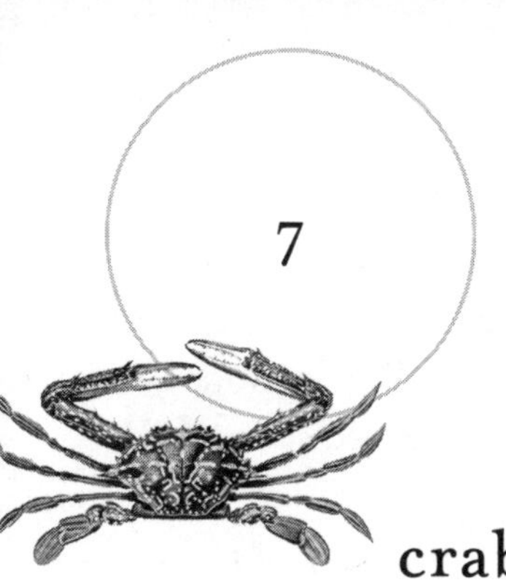

7

crab and clam treasures

The hermit crab was carrying a pen. Holding it tight between pincers, he was dragging one end through sand, making quite slow time back to his nest of shells and sea grape leaves under the bushes. Hermits are thrifty collectors, and with every intentional step, it was clear that this resourceful little scavenger was proud of his amazing find.

I marveled at the strangeness of the hermit holding a pen, poised as if attempting to scrawl a message in the sand. It was the same kind of black-ink ballpoint that was pressed inside the binding of the field notebook I carried, except that the hermit's pen had never been mine. From the salty and bedraggled look of it, the ballpoint had washed in from the sea. It had probably traveled a few hundred miles before being churned by waves upon the sandy shore, pushed by tides to the top of the wrack line in a mélange of shells, seaweed,

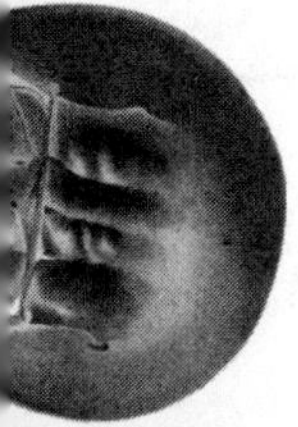

bits of broken sea fans, and various articles of pollution. From this amass of micro-wreckage, hermits prowled for their meals, mobile homes, and antiques. Tasty food was gobbled on the spot. New shells were claimed to replace the old, smaller shells they wore as they grew. Beguiling articles were painstakingly hauled across the beach to their nests among the tree roots.

I bent over the hermit. My shadow startled him, and he dropped the pen with a tiny thud and rapidly withdrew into his shell, concealed from sight. Without moving, I waited and watched. After a long moment, the hermit emerged, his antennae-like eyeballs scoping the sand for further signs of danger before landing on me, then his prize. He grasped the pen again and continued scooting his way along, faster than before, as if he feared theft over physical peril.

I watched his progress, debating with myself. One man's trash is a hermit crab's treasure, and clearly, the hermit desired to keep his newfound pen. He was working so hard to have it for his very own! But I knew the pen was trash that polluted the beach. While I didn't want to upset my new little friend, I also had noticed that the pen had been banged up quite a bit and, still being full of ink, had the potential to leak or explode. I pictured the hermit arriving home among the shells and leaves, joyously putting the prized pen on display among his most coveted possessions. I pictured a future night in which, spontaneously, the pen would seep toxic black ink, darkening sand, tainting the nest, and potentially hurting the hermit. I could imagine the bewildered crab trying to recover and escape. No, it wasn't worth any risk. The hermit might not know it, but the pen was potentially a danger.

I leaned over the hermit again, hoping he would drop the pen as before. But my close proximity only caused him to quicken his pace. "That isn't good for you," I explained. "It may seem nice now, but it will only cause you problems, so I'm going to take it away." I reached out with my hand.

The hermit was now scurrying faster over miniature sand hills than I had thought possible, with the incredible burden of the pen bending his pincer arm back in an impossible angle. He was trying to outrun me. Easily, I grabbed the opposite end of the pen with two fingers, gently beginning to pull it from his grasp. The hermit turned around, dug his spindly little legs into sand, grasped with both pincers, and engaged in what he perceived was the mightiest stand of tug-of-war. I chuckled while feeling pity for the little critter, so protective of this trash. After a brief moment of struggle, he began to slide through the sand in the pen's wake as I pulled. He would not let go; this would not do.

Reaching out my free hand, I tapped the back of his shell. Immediately, he let go of the pen and withdrew. Then, realizing he'd been duped, he rapidly emerged from his shell to stare at me with glazed eyes, utterly dismayed.

"I'm sorry, hermit. It's not good for you." He stood frozen by confusion and dejection, then slowly, turned and crept back to his nest.

The pen sat untouched on my desk for several weeks, grimy with salt and sand. The more I looked at it, the more I realized that so very often, I am like that hermit crab. I fight to keep things that aren't good for me. I have a tendency to place value on things that are worthless. I frequently hold on to cumbersome things I do not need to carry.

As with hermit crabs, it is normal for us to be grabby, snatching at things we want. "I want to be a marine biologist. I want to go to this school. I want a boat. . . ." Dreams for the future start with what we desire now, and it's not a bad thing to chase after them, so long as they are in line with God's Word and will. However, problems arise when we want the dream more than God; when we grab while disregarding what he says is for our best.

We have permission to plan but are counseled to do so with God at the heart of our decision-making. While our participation, perseverance, and hard work may be required, it is he, not us, who determines whether plans come to fruition.

> We can make our own plans,
> but the LORD gives the right answer.
> People may be pure in their own eyes,
> but the LORD examines their motives.
> Commit your actions to the LORD,
> and your plans will succeed. . . .
> We can make our plans,
> but the LORD determines our steps.
>
> PROVERBS 16:1-3, 9

Like a pen for a hermit crab or forbidden fruit for Adam and Eve, God's creatures sometimes desire things that he knows will cause problems. Indeed, we have free will to choose for ourselves. But sometimes when we refuse to let go of a problematic desire or dream, God might choose to take it away. Pens will do no good for hermit crabs. Dreams unaligned with God's perfect will do no good for us. If we trust him, relinquishing the grip on our plans will come more naturally. If we believe he is who he says, the loss of any dream will lead us nearer to his heart.

When I was younger, I placed high value on prestige. The better the school's reputation, the higher the job's qualifications, and the longer the person's credentials, the more impressed I was. It was only natural then that I funneled academic reputation, professional qualifications, and personal credentials into my dreams. God called me to my field of study and provided my vocation, but for a long time I let achievement be my primary driver, instead of simply aspiring to please and serve the God I loved. I was successful by nearly every worldly standard, but one accomplishment continued to elude me. Though it wasn't a requirement in my master's program, I was disappointed that I could never seem to get a research paper published. I tried multiple times with several people on many projects, but it never worked out. And it drove me a little mad because publications are to scientists what military medals are to veterans—important commemorative symbols of significant accomplishment.

It wasn't until many failed attempts and tearful laments later that I realized God's grace is sufficient for me and that his power is made perfect in my weakness.[1] If God wanted me to get a publication, it would happen. But it wasn't happening because, while I desperately wanted one and had worked hard for it, I didn't need it. If I had, I would have achieved it because God is with me and provides for all my needs.[2] I was laboring after something that, although ultimately good and well-intended, wasn't what God had for me at that time. I needed to relinquish my want in favor of his will. I let the research publication go. I opened my hands and grabbed on to God instead, the one who deeply knows the desires of my heart and (when I have my priorities straight) is himself the ultimate and only desire of my heart.[3]

Perhaps our greatest comfort in the wake of rerouted dreams or canceled plans is the discovery of what is meaningful. The book of Ecclesiastes recounts a wise man's pursuit of this. In what is frequently considered a highly depressing series of personal introspections, the wise man articulates, "I observed everything going on under the sun, and really, it is all meaningless—like chasing the wind" (Ecclesiastes 1:14). He concludes that the world, knowledge, pleasures, labor, time, relationships, wealth, life, and death are all meaningless.

However, careful reading of the whole book leads to a curious conclusion. While all things are meaningless, they are only so when they are apart from God. It is God who informs meaning. It is he who determines the value of who we are and what we do. It is he who distinguishes what is worthwhile and worthy.

> Follow the ways of your heart
> and whatever your eyes see,
> but know that for all these things
> God will bring you into judgment.
> So then, banish anxiety from your heart
> and cast off the troubles of your body,
> for youth and vigor are meaningless.
> Remember your Creator.
>
> ECCLESIASTES 11:9–12:1, NIV

The author encourages us to achieve things, live life, and follow dreams so long as we recognize that God is at the center of the story. We are not. Anything we do apart from God is meaningless. This is why he must be at the heart of what we desire and plan. He gives our dreams meaning.

Besides being a metaphor for things that are ultimately meaningless apart from God, the hermit crab's pen also reminds me of our burdens. The way that little hermit chose to drag a pen five times his own length and heavier than his weight is not an image I will soon forget. We, too, have the habit of hauling cumbersome things we were not intended to carry (or at least not carry for long). These burdens come in all sorts of colors, shapes, and sizes, often custom-made to the person dragging them along behind.

Some common "pens" we carry include anxiety, anger, and bitterness. These are things God longs to take from us. If we are anxious, he tells us to cast all our cares on him because he cares for us.[4] He offers to catch and carry our burden, if only we will give it up.[5] If we are angry, he reminds us of his own character—that he is "merciful and gracious, slow to anger and abounding in steadfast love and faithfulness" (Psalm 86:15, RSV). He enlightens our darkness, arming us with strength and making our way perfect, setting us on high places and teaching our hands how to fight in a way that honors him.[6] If we are bitter, he warns us of its toxicity, which will only spring up and cause trouble, defiling all it touches.[7] He enables us to put all bitterness and anger away from us.[8] Like hermit crabs carrying dangerous and unnecessary loads all the way home, we need to release our burdens to the only one who is equipped to carry them.

In addition to the burdens we choose to carry, we may be saddled with burdens we did not take up willingly. Sometimes they are thrust upon us, and try as we might, we cannot remove their weight. In many instances, we may beg for them to be lifted. Some of these burdens come in the form of illness, trauma, and grief. These are not chosen, and unfortunately, we cannot simply give them up and be done with them. Certainly, we can continue to press into God, casting all our cares on him, but that will not necessarily alleviate all our present pain in the way we may hope or ask. Such burdens are not comparable to the one in the hermit crab's story, but our response to them may be compared to that of a clam.

When I first saw the giant clam bed while snorkeling in the pristine waters off Australia's Orpheus Island, my eyes nearly bugged out of my head. Not only was the clam bed widespread across the reef, the clams themselves were ginormous! Masses of *Tridacna gigas*, the world's largest living bivalves, stood upright in the sandy substrate, their lengths larger than my arm span and shells thicker than my hand, encrusted with a mosaic of algae, sponges, and tiny corals. Their thick-coated mantles protruded like massive rainbow-painted lips. Purple and green, orange and blue, each mantle was a unique pigment composed from symbiotic algae, shimmering with tiny iridescent rings that dotted the smooth yet crinkly flesh. On each clam, a siphon, the mouthlike tube through which the clam ejects water, was asymmetrically positioned inside the mantle. On the opposite side of the mantle, a secondary crevasse-shaped siphon drew water into the clam. By sucking a large amount of water in through one siphon and expelling it through the other using powerful internal muscles, the clams could close themselves.

I swam over a clam, admiring its iridescent blue lips. Having done my research beforehand, I knew that despite legends and colloquial nomenclature, these clams were not man-eaters. Like most wild animals, they were likely to be far more afraid of humans than we were of them. I took a deep breath through my snorkel and dove down, my gloved hand outstretched. Gently, I touched the clam's blue mantle with my finger. The flesh retracted slightly into the shell, as if ticklish, before slowly unfurling once more. Delighted, I touched the mantle again, this time with my whole hand. The ticklish clam lurched its mantle inward, the sides of the shell jolted toward each other with the noise of a scratchy cough, and a cold jet of water shot from a siphon into my face. The shell closed quickly at first, then slowed as the two halves neared before tenderly touching together. Shut tight, only the faintest outline of colorful mantle could be seen between the wavy shell edges.

For the rest of that afternoon, I could be seen either diving down to tickle giant clams or floating on the surface, giggling into my snorkel. When the clams were startled or uncomfortable, they consistently hid themselves away, withdrawing their soft parts into the protection of their hard outer shells. Mantle retracting, siphon jetting, and shell closing. This ability is a unique defense mechanism, designed to protect them from danger (and apparently, tickle tortures).[9] But besides being a unique physiological phenomenon that is immensely enjoyable to the interactive snorkeler, this ability can serve as a metaphor for the way in which we humans respond to burdens that won't lift.

When things that we have no control over crop up in our lives—when illness, trouble, trauma, or grief appear—it is natural to respond much like a giant clam. We retract into ourselves, we push back from the things creating pain, and we close out the world and maybe even God. It is not my intention to be critical of this reaction, but the simple fact of the matter is that while a clam may initially close in order to survive, a clam that stays closed will eventually die. Unless it opens, the clam cannot get its necessary nutrition from filter feeding, sucking plankton in through a siphon and expelling subsequent waste products.[10] Unless it reopens, the clam also cannot expose the symbiotic algae living in its mantle to sunlight so that it can photosynthesize. Without this process, the symbiotic algae will die, and the food it would have provided for the clam in exchange for its home will be lost.

Not only that, but the beautiful pigmentation of the mantle, the iridescent blue-ringed patterns created by these algae, will fade. The clams will lose their color and their lives. It is okay for a clam to close for a time, but it cannot stay that way.

Similarly, when we face unchosen burdens that remain heavy upon us, an understandable reaction is to close up like a giant clam. At best, it may provide a moment of respite to turn away from all that doesn't ultimately matter and press all the more intentionally into the heart of our loving God. But if we close ourselves off from him and stay that way, we will not survive for long. Just as individual clams may choose to reopen more slowly or quickly, people choose to open themselves again based on their own capacity and time frame. But eventually, we need to reopen. We need to decide whether we will continue to trust God.

Discomfort, sorrow, and undesired change provoke a similar reaction. Interestingly, it is through these experiences that God often grows us. Remembering this, "we can rejoice, too, when we run into problems and trials, for we know that they help us develop endurance. And endurance develops strength of character, and character strengthens our confident hope of salvation. And this hope will not lead to disappointment. For we know how dearly God loves us, because he has given us the Holy Spirit to fill our hearts with his love" (Romans 5:3-5). We need to consider our response when God's plan differs from our own, when we are perhaps called to lay aside a specific dream or plan. Do we close up like a ticklish clam? Or do we continue to trust a most trustworthy God? Do we batten down the hatches and give up living, determining to just wait it out until things finally get better? Or do we continue to seek for meaning in the one place we will most certainly find it?

Whether our personal circumstances align more with the story of the hermit crab or giant clam, there is one thing we can know with certainty. Both stories represent opportunities to reorient our perspective. When a "pen," plan, or dream is taken from us, or when an unchosen, unmoving burden lies heavy on us, we have the opportunity to turn away from what we have lost and how we feel, toward the Creator of true treasure.

If we have confidence in God's character, we can recognize that the removal of something we desired will be for our ultimate best. The pen will not take up space in our nest, so what true treasures might it be filled with? In the void left behind, there is now freedom for good things from God. "So then, banish anxiety from your heart and cast off the troubles of your body, for youth and vigor are meaningless" (Ecclesiastes 11:10, NIV). We should not fret over the loss of something God deems has no value for us. We should instead seek after what is truly valuable, God himself. We reorient our dreams around him, seeking his will over all things. Humbled in the wake of our rattled plans and tattered dreams, no longer within the scope of future reality, we cast our anxiety on him because he cares for us, remembering that he will lift us up in due time.[11]

Unlike us, God is not weighed down by burdens. In fact, "God's weakness is stronger than the greatest of human strength" (1 Corinthians 1:25). Because of his power, we need not carry burdens alone. We press into God, believing he will use suffering to produce perseverance, character, and hope within us.[12] We recognize that even in nature, discomfort and change are symptoms of unfolding beauty. A caterpillar must wrap itself in a cocoon of tightly spun thread and wait in confinement in order to grow gloriously colored wings and take flight. A seed must be buried in the moist darkness of soil before it can burst from the ground and blossom into beautiful flowers. A single gritty speck of sand wedged uncomfortably into the flesh of a bivalve (such as a clam or oyster) may eventually yield a large, iridescent pearl. The same is true for us.

Focusing on the sand-turned-pearl analogy for a moment, the suffering of the clam becomes evident. The grain of sand grits across their mantle, becoming wedged in their bodies where they secrete mucus around it to soften friction. This causes the grain to grow, creating even more discomfort. In a spiral of uncomfortable secretion and subsequent growth, the grain becomes large, smooth, iridescently brilliant. There is beauty born from the change, growth, and discomfort of the burden, and one of the main ingredients in the transformation from speck to pearl is time.

> For everything there is a season,
>   a time for every activity under heaven.
> A time to be born and a time to die.
>   A time to plant and a time to harvest.
> A time to kill and a time to heal.
>   A time to tear down and a time to build up.
> A time to cry and a time to laugh.
>   A time to grieve and a time to dance.
> A time to scatter stones and a time to gather stones.
>   A time to embrace and a time to turn away.
>
> ECCLESIASTES 3:1-5

There is a time for caterpillar and butterfly, seed and flower, clam and pearl.

Many years ago, a local fisherman's anchor snagged on a giant clam off the coast of the Philippines. The fisherman swam down, unhooked the anchor, and discovered that the clam contained a treasure. It was a giant white pearl, over two feet long and weighing seventy-five pounds. He hauled the treasure home and, not initially comprehending its full value, stored it under his bed for ten years as a good luck charm. When eventually shown to the world, the pearl was valued at one hundred million dollars.[13] The pearl's formation had required years of pain and secretion and effort on the part of the clam to eventually create the seventy-five-pound growth of beauty and value within its body.

God can work through discomfort and pain to create incredible things. But will we be open to that? There are times to acknowledge that the grain of sand is uncomfortable, and there are times to believe in the pearl it could become. We may not always understand; we rarely do. We may close up into protective shells for a moment, but as there is a time for everything, so we must open again, choosing to trust God.

> God has made everything beautiful for its own time. He has planted eternity in the human heart, but even so, people cannot see the whole scope of God's work from beginning to end. . . . I know that whatever God does is final. Nothing can be added to it or taken from it.
>
> ECCLESIASTES 3:11, 14

We release the pen and trust God. We wait for the good things he has for us, casting our cares on him, because he can carry them and he cares for us. We hold the grain of sand in our discomfort and press into Jesus. We remember that he alone has the power to make treasured pearls within us and to bring meaning to our lives. He might take treasures so that he might make treasures of far greater value than those belonging to crabs and clams.

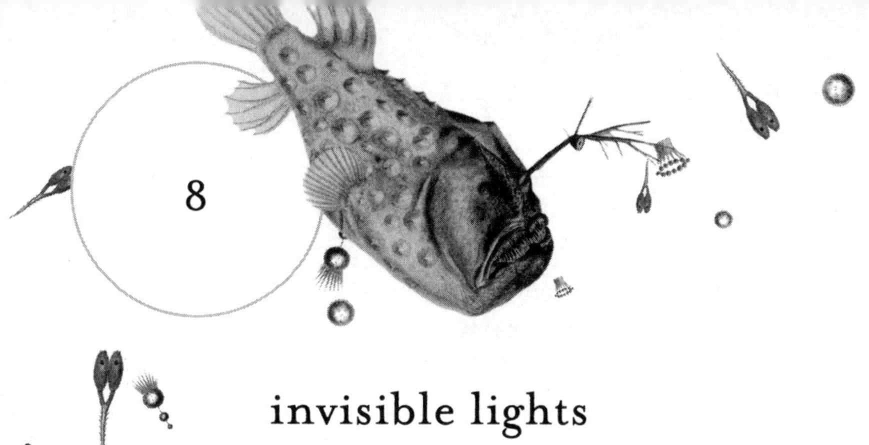

8

invisible lights

In the chill of late spring, the scent of cherry blossoms wafted through the tall grasses where my coworkers and I lay with backs pressed into the soil, eyes staring upward into the night sky. We scanned the blackness glimmering with silver pinholes of light. In the distance a fuzzy red orb, maybe Jupiter, flickered. I tried to focus on its glow, but it was hard to see. Earlier that day, our whole team had been laid off due to a global pandemic. As rumors, fears, and symptoms of the coronavirus spread across the country, our jobless team was quarantined to the isolated field station where we now lay, spread under stars, confiding our fears and wishing to glimpse an unknown future. We struggled to focus our minds on hope, just as we strained our eyes toward the stars.

I was marveling at the depth of the celestial bodies glimmering like bioluminescent plankton in a dark sea and imagining what astrophysicists preach, that the universe is steadily expanding, galaxies unfolding like tentacles of a coral polyp. I knew there were countless glories of light beyond what my eyes could see and I wanted to lie there in the darkness forever, looking at the sky that made me ask big questions about life, meaning, the future, and God. Turns out, I wasn't the only one asking.

One of my coworkers was an intelligent gal with short, wavy hair, now fanned out in the fresh spring grass. We had bonded over our common interest in the aquarium hobby, and I thought of her as inherently practical and bravely adventurous. After lying for a while in silence, she abruptly sat up and spoke words that surprised me: "I can't look at it too long. It's so terrifying."

"What's terrifying?"

"That," she said, pointing up at the starry sky.

I paused to consider, turning my eyes upward again. There was the thin presence of an oncoming new moon, a place of luminescent sand pocketed with craters and marked with the light gravitational tread of astronauts. Strands of our Milky Way galaxy swirled with splendor, embedded with the pulsing light of planets and sparkling stars. These stars are born, burn, and die as luminous balls of gas held together by their own gravity, their entire lives defined by light. It was amazing and almost comforting to feel my own smallness under the sky, reminiscent of standing barefoot at the edge of the roaring ocean, where waves crashed and ran up the beach to gently touch my toes. My life and its myriad of problems and questions was so small by comparison.

"It just goes on and on . . . and we don't know what all is out there, even though we're trying to find out. It's so big and unknown! And that scares me."

She was right. There is so much we don't know. Somewhere

beyond our solar system lies the Bubble Nebula, where heat from a central star warms a nearby molecular cloud, causing the cloud to coagulate and form a transparent, glowing bubble shape, wrapped around the young star. Farther beyond lies the Pillars of Creation, where cosmic fingers of interstellar gas birth stars in dusty nurseries. So much beyond our reach physically, but also intellectually. As famed astrophysicist Neil deGrasse Tyson has written, "What we know is that the matter we have come to love in the universe—the stuff of stars, planets, and life—is only a light frosting on the cosmic cake, modest buoys afloat in a vast cosmic ocean of something that looks like nothing."[1] As with the ocean, there is an awful lot of something out there. As technology advances in the field of astrophysics, as in most branches of science, our knowledge has the potential to expand too. But even as our understanding of the universe has grown due to theoretical and applied physics and an impressive amount of cosmic research, there are admittedly some things out there that may quite literally be beyond us.

As my coworker and I stared up into the vast cosmic ocean, I realized just how different our perspectives were. We saw the same night sky, and yet the very thing that bred fear in her created wonderment in me. Her perspective was steeped in terror. Mine was that of open-mouthed awe. As we lay hypnotized under the weightless light of "something that looks like nothing," I let my mind drift into the darkened corners of all the unseen things that exist, like the deepest stretches of the ocean floor.

Ninety-five percent of space on earth is perpetually dark. At a depth of 650 feet, the vast majority of light can no longer penetrate the deep sea.[2] Few have glimpsed this eternal night. Those who have scuba dived to the point where colors vanish without a trace or who have engaged in night diving have consciously and curiously immersed themselves in a territory entirely foreign to their own, staring into a void of blackness.

The deeper you go, the darker and more pressurized your surroundings become, which is why modern deep-sea exploration entails the use of robotic underwater vehicles. These vehicles send back images and videos of an otherworldly space. Underwater mountains rise and fall, studded by the constant out-jetting of hydrothermal vents, cracks in the seafloor that release magma-heated water and minerals. These chimneylike vents support a peculiar and diverse ecosystem where bacteria thrive, tube worms burrow, and octopuses frolic. In the valleys between vents, crabs congregate to forage on whatever nutrients can be found—a whale fall, perhaps. After being scavenged by seabirds and sharks, a whale carcass sinks, eventually settling as a heap of nutrient-rich feast for the deep-sea community and enriching fertilizer for the seafloor. Tiny flakes of marine snow drift down from the surface—collections of cobweb-like mucus from ctenophores, leftover bits of yesterday's sea lice, and other dust bunny shreds of organic material. While these scenes take place mostly in darkness, the deep sea is not altogether without light. Speckles, streaks, and pulses of color glow in the darkest spaces. The poets call it magic, but scientists named it *bioluminescence.*

Bioluminescence is the production and emission of light by living organisms. Although we may not witness bioluminescence very often in our daily lives, it is a relatively common biological phenomenon. Light-producing terrestrial organisms include fireflies and other insects, as well as some bacteria and fungi. In the deep sea, bioluminescence is primarily used for communication, which plays a particularly important role in courtship and mate selection. For example, as flurries of bioluminescent copepods (bug-like plankton that look like the villain from the TV show *SpongeBob SquarePants*) swim through the darkness, they may be saying some version of, “Hey there, good-looking, you really light up my world!” In other instances, bioluminescence can serve to enhance a fish’s hunting endeavors, aiding in their overall survival. Ambush predator anglerfish (as seen in the movie *Finding Nemo*) use an overhanging bioluminescent lure, called an illicium, to tempt smaller fish into their crescent-shaped mouths filled with bouquets of long, angled teeth.[3]

Organisms that bioluminesce usually have a special part of their bodies where the light-producing chemical reaction occurs. In the reaction, two main ingredients are involved: the molecule luciferin (named for a Latin word that translates as “light-bringer”) and the enzyme luciferase. When luciferin and luciferase come together with energy and a dash of oxygen, luciferase undergoes a shape change from “open” to “closed,” where it wraps around luciferin. These molecules, now hugging, become excited and release photons in the form of light.[4] While this is the basic recipe for bioluminescence, there are a variety of different luciferins, luciferases, and photoproteins (luciferase replacements) involved in versions of the reaction, as well as different anatomical organs where the reaction takes place. To use an analogy, there are multiple ways of baking a delicious cake. One recipe might call for a little more flour and another recipe might replace eggs with applesauce, but the process usually takes place in a kitchen and always involves putting batter into an oven, and the end result, although not altogether identical, will be cake.

As with deep space, there is so much we do not know. Despite our continual underwater exploration and the illumination provided by bioluminescence, much of the deep-sea world remains shrouded in darkness. But from the perspective of a resident fish, could things look different? Perhaps fish have a different way of viewing the world than we do.

In the ocean, most fish have vision adapted for the surrounding darkness. Heightened light sensitivity allows their eyes to recognize near-untraceable amounts of light as normal, picking up wavelengths that would be undetectable to us. Other fish have adapted their bodies to accommodate their vision needs, having large, bulbous eye shapes or lenses angled toward the surface to detect silhouettes.[5] And as we've seen, some fish produce their own light by means of bioluminescence.

Certain fish have yet another means of using light to their advantage. Consider the guppy, native to the fresh waters of South America but ubiquitous in fish tanks all over the world. When seeking mates, females are especially attracted to males with the handsomest colors—most of which are invisible to the human eye.[6] Humans might look at a fish and see only two or three colors, but to the other fish, a rainbow of color appears.

This is tied to the electromagnetic spectrum, which requires a brief tour into the realm of chemistry and physics. Hundreds of colors exist invisibly in rays of light around us, bouncing off water and reflecting off our own skin. Thousands of shades of color are indistinguishable to our eyesight, but they still exist and, if we're using the right equipment, are detectable.

Everything in the world is made of atoms, which function as tiny interacting building blocks. There are different kinds of atoms, based primarily on the number of protons (positively charged particles) they contain. You would probably best recognize these different atoms, called elements, from a figure of the periodic table, found in the front of every decent chemistry book. When multiple elements fit together to build a larger sort of building block, it's called a molecule. These work somewhat like Legos, building bigger things from smaller ones. And what's perhaps most remarkable is that "every element, every molecule—no matter where it exists in the universe—absorbs, emits, reflects, and scatters light in a unique way."[7] When molecules interact, energy is almost always involved. Typically, energy is either emitted or absorbed, and this energy can take the form of light.

Energy functions much like ocean waves, which may be small or large. If we were to line up the different kinds of energy wavelengths and order them from smallest to largest, they would exist on a continuum, called the electromagnetic spectrum. The smallest waves are called gamma rays, and as they increase in size, they morph into X-rays, followed by ultraviolet light. The waves are now what we might consider medium in size, and this is where the visible spectrum of light exists. All the colors we humans can see, from violet to red, exist in this tiny section of the electromagnetic spectrum. As waves get larger past visible light, there is infrared light, then microwaves, and finally radio waves.

The visible light spectrum that humans can see is only a tiny fraction of the entire electromagnetic spectrum. Every wavelength that is smaller than visible light is invisible to us; every larger wavelength is invisible too. But just because we can't see something doesn't mean that other creatures cannot.

For example, some kinds of fish are able to see ultraviolet light, using colors invisible to us to communicate, find food, and choose mates—like those guppies. Two stripe damselfish use the ultraviolet coloration of their bodies to flash signals to surrounding fish in the presence of predators, alerting them to danger. Brown trout can see ultraviolet in their juvenile life stage, where this special vision helps them spot and catch delicious zooplankton in shallow, sun-saturated water.[8]

Vision impacts our perspective, so when we switch from visible to ultraviolet light, our understanding of reality transforms. The same could be said for the conversation that was had under the starry sky that late cherry blossom–filled night. My coworker saw that there is so much we don't know, that there are black spaces between the stars, and so much darkness. And that was terrifying. When I gazed up at the stars, I saw something quite different. The heavy sky was not merely void and colorless but rather pointed my mind toward infinite and eternal hope. The same sky that brought her terror brought me wonder and comfort. "When I look at the stars, I see someone else."[9]

This difference of perspective exists between those who do not yet know and love God and those who do, because there is a difference in what we can see. An unfolding display becomes accessible when we have learned to know the Creator and begin to see through his eyes rather than our own. What was once invisible becomes visible. Our visual portion of the electromagnetic spectrum expands as we gain the spiritual equivalent of fish eyes. We are able to "fix our eyes not on what is seen, but on what is unseen, since what is seen is temporary, but what is unseen is eternal" (2 Corinthians 4:18, NIV).

Although humans cannot see most of the electromagnetic spectrum beyond visible light, pretty much every part of it is used by astrophysicists to study, research, and image space. I have long possessed a weakness of procrastinating by looking up NASA images from the Hubble and James Webb Space Telescopes.

These powerful instruments live in space, providing pictures of neighboring galaxies, exploding supernovas, and distant globular clusters. Just as fish eyes and human eyes pick up different colors in the electromagnetic spectrum, different kinds of telescopes are designed to pick up on different wavelengths of light. Orbiting some 320 miles above the earth's surface, the famed Hubble telescope sends pictures using mostly visible light, plus a bit of infrared and ultraviolet radiation.[10] And about one million miles away from earth, the newer James Webb telescope orbits the sun, taking images in infrared wavelengths.[11] Both telescopes make visible the wavelengths of light that are invisible to the human eye, providing images of some of the most phenomenal reaches of our universe. Simply put, they make the unknown a little more knowable.

Just as a well-aimed telescope can clarify that what appears to be empty space is actually filled with heavenly glory, science has the power to uncover bits and pieces of the workmanship of God. Science enlightens the hearts of those who love its Creator. Christians may gaze at the night sky with wonder in our eyes, but scientists point a telescope at the black spaces between lights and attempt to look beyond the visible color spectrum. By and large an enthused and nerdy type, we scientists seek to find the light of understanding in even the darkest of spaces. We are like fish that live in deep waters of the ocean's twilight zone, able to see in the dark with specially adapted eyes. All that to say, we see that there is more to discover! The universe fascinates us, and we long to hear its stories. For these reasons, having conversations with scientists and engaging with the science itself has the potential to deepen a Christian's understanding of the created world, thereby revealing more of the character of God.

But science points its telescopes into the dark with excited and frequently quivering hands. Its most vexing question is, Why is there something instead of nothing? Like my intelligent coworker who was afraid to look at the night sky too long for fear of the answers she lacked, some scientists would rather know things at face value without questioning their meaning or purpose. They prefer to see light without asking why it exists, because the immensity and intensity of a nearly unfathomable creation makes the concept of facing something or someone beyond that feel perilous. This is where Christians are meant to shine in the darkness. This is where faith transforms perspective. This is where people who love God have the potential to describe the reality of creation in terms beyond what is visible.

We can all recognize that there is much we do not know. There is more to the electromagnetic spectrum than visible light—more out there than the lenses of our telescopes or robotic underwater vehicles can presently detect. And certainly, there is more to our lives than uncertainty. There is more to be known of the Creator than what we currently know. In a way, both scientists and Christians can be invisible-light goggles to the other, expanding the spectrum of our bare eyes' ability to take in light from darkened places.

A cold, vast universe feels foreboding until informed by the reality of God's loving character. The dark, bizarre ecology of the deep sea seems terrifying until we recognize who created it. What initially filled one with fear now results in open-mouthed wonder, as "perfect love drives out fear" (1 John 4:18, NIV). Why are there stars in space and bioluminescence in the sea? Perhaps those lights were put there for us to see. A bland, face-value acceptance of creation is overpowered by deeper understanding of how this colorful, wildly amazing existence forms and functions. And our minds spin with wonder and delight as we take in churning galaxies, the unceasing dances of glowing fish, and every potential wavelength of light, all of which reflect back to us glory beyond glory.

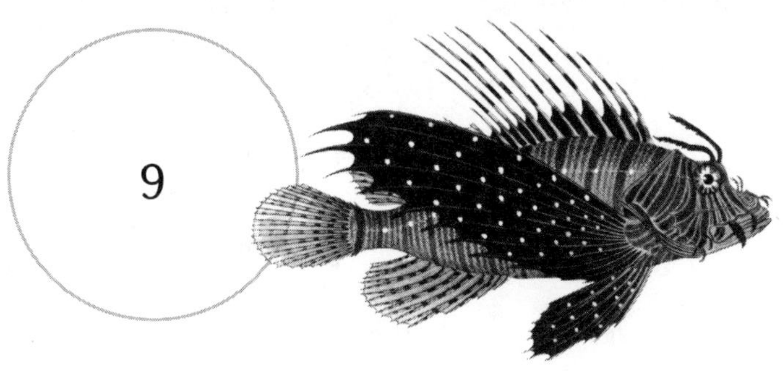

9

buddy breathing

My family used to make annual pilgrimages to the Hiawatha Trail, a fifteen-mile mountain bike ride on an old railway line that takes you over seven trestle bridges and through nine tunnels along the border between Idaho and Montana. Needless to say, with its alpine expanses and gorgeous views of the Bitterroot Mountains, the ride is worth doing. But perhaps the most notorious aspect of the Hiawatha is the mile-long tunnel at the start of the ride, which, to a little kid, is the stuff of bad dreams.

The mile-long tunnel (actually named St. Paul Pass or Taft Tunnel) looks like a giant void in the side of a mountain. It runs under the state line and despite its nickname is truly 1.66 miles long.[1] Long enough that you can't see any light at the other end. As you pedal into it, you see only inky darkness.

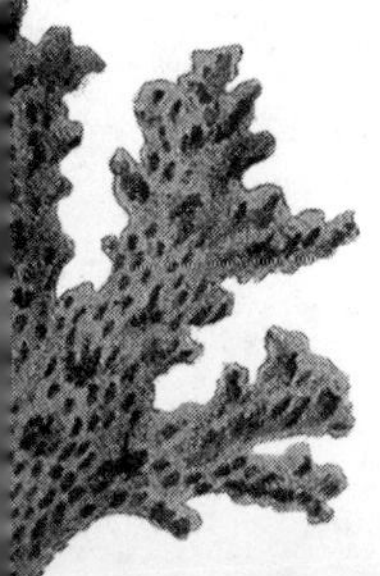

Inside, it feels more like you're about to run into a giant piece of black construction paper, but instead you just keep reaching void space. If you're unlucky, you hit a wall of the mountain you're under.

There is no safe way to ride a bike through the tunnel without a light. Ideally, you would have three lights: a headlamp, a flashlight on the handlebars, and a light attached to the back of the seat so that nobody rides into you. You move slowly, keeping track of forward speed by the motion of your feet on the pedals rather than by the passing of your surroundings. As you ride, your eyes adjust to the blackness, but that doesn't mean you can see any better. It just means that somehow your body learns to stop straining for light that doesn't exist. There's only black, and if you keep a super keen eye out and the flashlight beam angles just right, you may notice the historical plaque of the railway line mounted to the wall at the halfway point.

The most unnerving part of the tunnel is the middle. If you take a moment to turn off your lights, you realize you can't see light at either end of it. Where you came from is black, where you're going is black. It wraps around you, thick as a heavy felt blanket over your face. Dripping sounds coming from the oozing groundwater above, which runs down the walls and splashes into little mucky troughs on the sides, make it extra eerie.

The first time my family rode the Hiawatha, my brother and I were young and got scared. I don't remember how worked up we were, but I do remember how overwhelming the dark felt. I remember wanting to turn back. And I remember the response of my kind parents; they told us to sing. One of them started singing "Amazing Grace" and told us to join in. We sang hymns, VeggieTales songs, and a rousing new version of "She'll Be Comin'

Round the Mountain." It helped. Partly because it was distracting. But I think the greater reason was because our voices singing hope and joy into the void made it seem less dark.

I hardly know how to describe the feeling of when light finally appears on the other side. At first you think your eyes are fooling you. Maybe it's just a speck on your eyelashes. Then somebody else says they think they see it and you dare to hope. The light gradually grows, but it emerges so slowly that you cannot tell unless you glance away for several minutes and then look back. And then, after a much longer time than you expected, the brightness emerges and swallows your field of view.

You come shooting out into sunlight. Your senses are so dazzled and disoriented that, if you don't immediately stop and dismount, you're practically guaranteed a crash. You're a mole squinting into the world, the white slowly fading to the majestic greens and blues and browns of the scenic beauty surrounding you. If you can make it through that tunnel, you can easily deal with any other challenges the rest of the gentle Hiawatha ride may throw at you. A trestle bridge some one hundred feet above the treetops? Sounds fun. A moose standing in the middle of the trail? So cool. Thirteen-plus miles still ahead? Let's go.

I've seen solar eclipses and dived to great depths, but no other experience has been so literally dark as the mile-long tunnel. Riding through it taught me numerous things, like the importance of planning ahead and maintaining self-awareness to stay safe, and the ability to find courage in the face of fear by the simple practice of singing. The greatest thing it taught me was that dark places exist where you cannot see light. Sometimes, in some places, you cannot even see God. When that happens, you have to decide how you will respond.

Any diver worth their salt has struggled with fear at some point. Some have gone for hundreds of peaceful dives only to encounter "that" fateful dive, the one where equipment malfunctions or they get lost underwater or a buddy runs out of air. Some divers are scared before they even hit the water. Their hands are shaky while gearing up as they fight off predive anxiety. In diving, fear is a Goldilocks concept. Some is healthy. It keeps you smart, aware, and humble. No fear makes you cocky, and too much fear is debilitating. Both can get you killed.

Because of this, one of the core principles emphasized in any scuba diving course is the concept of remaining calm. "Keep breathing" is the common refrain. Dive instructors are trained to identify early signs of panic in newbies. In more advanced training, you're encouraged to think through worst-case scenarios, how you could best handle them, and challenges in conducting those responses. You become a master of flexing contingency plans and rerouting emotions to keep them out of the way of your rational brain. Because when things go wrong underwater, the line between controlled fear and all-out panic becomes decidedly thin. How a diver responds to fear is frequently the deciding factor for survival, because the unfortunate reality is that a diver who panics can easily become a dead diver.

Similar to the apprehension I felt for the Hiawatha's mile-long tunnel, in my early days of diving, I had to fight horrible bouts of predive jitters. I never struggled with seasickness outside of a diving context, but geared up with a diving cylinder strapped to my back, I was often off the stern "feeding the fish," as they call it. And I was jumpy in the water. When a big, well-intentioned fish unexpectedly swam past, I'd feel my heart rate skyrocket. It was normal for dive buddies to check up on me excessively during dives, constantly giving me the "okay" signal; they could see how hard I was breathing by the steady river of enormous flat-bottomed bubbles I exhaled.

I had just started working as a coral biologist at Dry Tortugas, exponentially leveling up in my experience as a professional diver. Trying to prove myself in a competitive field with highly accomplished colleagues, I'd been reticent to share my "normal" fears, but the head dive officer had noticed my jitters. He was a burly Navy veteran with jam-jar glasses, a loud personality, and an even louder voice. He wore an orange beanie to the dive boat, hot pink fins in the water, and on surface intervals he ate beans straight out of the can without a fork. He was unpredictably frightening and likable, the sort of guy you want on your side.

My first weeks on the job were exhausting but happy. Group training pushed my body to the limit, and I felt the strain of isolation that happens when you are alone with your thoughts and feelings in an environment that requires professional behavior but breeds the casual attitude expected of beach bums and surfers. There were many tests—physical, mental, and emotional. I struggled to pass the single-breath-hold underwater swim where, without any gear, I swam seventy-five feet on a single lungful of air. In the process, I accidentally plowed headfirst through a school of massive moon jellies. My cheeks tingled with their stings for days after.

We did timed swims, distance swims, ditch and dons, and bailouts. Sometimes we all heaved ourselves from the water, gasping for breath like fish. Then there was the mental work. There were Navy dive tables to learn, scuba math to master, dive computers to understand, and underwater navigation to practice. We also had to learn about air compressors, gas analyses, rescue diving, safety procedures, vessel operations, and equipment repairs.

One of the last days of training was logistically challenging. A lot had gone sideways, and we had to rearrange and cancel dives. The stress was mounting in us all. It was windy, so the water was turbid and rough. Ten feet down looked more like forty, and forty looked much deeper. It was in these less-than-ideal conditions that we would perform the buddy breathing exercise.

Buddy breathing is an essential skill for a safety-conscious diver to master. It is the procedure to use when you or another diver are in an "out of air" scenario. In other words, if you or someone else can no longer breathe underwater, this exercise will save a life.

When practicing, you pretend to be out of air and breathe off your buddy's spare regulator. You take off your mask so that you can't see, while your partner grabs your forearm in a Roman handshake. Then it's your job to do nothing but be completely blind, stay calm, and keep breathing off your buddy's air supply while they adjust their and your buoyancy control devices (BCDs) to ascend back to the surface at a slow, steady, safe rate.

Though being "out of air" is just a drill, the sense of urgency is still extremely high. The rescuer needs to be attentive and knowledgeable because they guide their buddy through the whole process. They are responsible for two lives: their own and their buddy's. While they carry most of the cards as to what happens in a rescue scenario, the person being rescued carries the ultimate trump card—the choice to stay calm or panic.

In many ways, being rescued is the far harder role of the exercise to play. The person being rescued needs to trust their rescuer.

If they don't, a real safety incident could develop in an instant. Nervous divers are known for panicking, and panicking hard, in such scenarios. Even as a practice exercise, one wrong move could be perilous, one moment of panic could result in disaster. Hence, the head dive officer would hover next to us in the murky water at forty feet while we performed the exercise.

My assigned dive buddy that day was a strong gal with perfect beach-blonde hair, a quiet voice riddled with curse words, and a habit of impatient pacing. Although she was young, we both knew her list of diving credentials far surpassed my own. Together, we were about to do the most intense trust exercise either of us had yet experienced.

You and your dive buddy hover suspended at forty feet in gray-blue as particulates flood past your field of view. The head dive officer with pink fins gestures to you. You will pretend to be out of air first. He is watching. Begin.

You remove the regulator from your mouth and drop it, now no longer able to inhale, while steadily exhaling tiny bubbles.[2] You frantically signal to your buddy that you are out of air, by making a slicing motion across your neck with your whole hand, back and forth. They shoot toward you, spare reg extended. You insert their regulator, purge it, and breathe deeply. Sigh in relief. You are now connected to their gear by a hose extending from your mouth. While you hold the reg tight to your mouth with one hand and grab on to the straps of their BCD, they grab securely on to you. To be ripped apart by the current would leave one diver—you—without air. Your buddy checks in with the hand signal, "Are you okay?" You signal an okay sign back and proceed.

To simulate a potential hazardous situation, you have to take your mask off your face. You close your eyes as you feel the mask seal peel from your cheek, cold water flooding in. Completely filled, the mask comes off easily in your hands. Careful not to drop it, you slide the mask's strap up your arm to your shoulder, thankful all the while that you can feel your buddy still holding on to you. If you open your eyes, the cold salt stings them and everything is blurry blue. So you keep them closed.

The world is dark, and the gurgling blast of bubbles flooding across your face is echoed by the quieter gurgles from your partner, whose arm you cling to. You have to trust they've got you. And you know there's a venomous lionfish somewhere out there in the dark because you saw five nestled in a lost anchor just prior to removing your mask.

You're suspended in the water column, with no awareness of moving up or down. Your nose burns. The brine taste in your mouth is the least of your thoughts. Exhaling is alarmingly noisy. You feel your partner adjusting the BCD on your left shoulder. Keep calm, keep breathing.

Brine, bubbles, noise, burning.

Wait. What's that?

Your world gradually lightens. You feel the darkness slowly, painfully slowly, ease into lighter space. Do you kick your fins to go faster? You want to, but no. Your partner holds your right arm. You have to trust they will kick for both of you.

Gradual lightening. Bubbles catch in your nose. It's still cold and dark, but lighter each moment. You hope you're not just fooling yourself.

Lighter. Lighter. Farther to go. Surely almost there? Lighter. Where's the surface? Lighter. There . . . ? No. Your mind spins as you try to relax into the moment. Keep breathing.

Lighter. More water, more time than you'd hoped or expected, but all the while you see things lighten. Ten feet? Two? Fifteen? You don't know. Lighter.

A wave rushes over your head. You burst on the surface in full light, like coming out of a dark tunnel into dazzling splendor. Blue sky above, eyes open, reg spit out of mouth, you manually inflate your BCD. Your buddy watches and then signals to ask if you're okay.

You're both okay. You both trust. The exercise is complete, and while you look at each other, you think to yourself that there's probably a spiritual lesson in all of this.

It's easier to trust someone who has already proven themselves trustworthy. Who already has a track record of being there for you when you need them most. Who you know has integrity of character and the resolve to care for you.

That was one reason the buddy breathing exercise was so challenging and scary for both my buddy and me. We didn't know each other. Not well, not really. We knew enough about each other to know that we probably wouldn't be besties who wanted to have pizza nights and regular phone calls and hangouts on the weekend. I knew she paced when nervous. She knew I fed the fish. But we didn't really know what sort of person the other was. We didn't know if the other diver was a good partner. If she was trustworthy. If she would be capable of saving our life, should we ever need saving.

Buddy breathing changed that. There was a mutual revelation of abiding trust. We were now confident that the other was capable of remaining calm and rescuing another diver. I have since dived with people far more experienced, more skilled, more highly trained, and more decorated. But I have never felt as wholly safe diving as I felt with her. Because she was there for me when I needed it. When I was scared and in need of rescue, she was the one holding my arm, taking care of me, guiding me to the surface.

It is the same with God. It's easier to trust him when you really know him. It is easier to trust him when there's a track record of all the times he's been there for you—when he was holding on to you, taking care of you, guiding you to the surface. It's easier to trust him the next time you're scared underwater because you know he rescued you before. He can do it again. You aren't as likely to doubt his ability because you've seen him do great things. It doesn't ultimately matter that you can't see the light at the end of the tunnel. It doesn't matter so much that you're uncomfortable beyond belief. That you don't know when the misery and darkness will end. That you're terrified. You know there's light to be found, and he will get you there. It may be much further off and later in coming than you want it to be, but you know you'll reach it faster, better, and more safely if you trust him. He knows and sees what you do not. And so you press into the feel of his secure hold on you. You keep breathing and trust him.

Hours after the buddy breathing exercise, a lovely sunset lit up the sky as I watched the frigate birds rise and fall over Long Key. They're all points and angles when they fly, with silhouettes similar to bats. They fly together, slowly circling over the islands like a lazy cyclone. I walked along the windy shore, watching boats move in and out of the channel.

I got back to the team house shortly before the burly dive officer brought over a bottle of spiced rum, the nectar of pirates, scalawags, and divers. Well-meaning and generous, our officer intended this as a symbol of celebration. And we were commemorating more than just having completed training exercises. The dive officer tossed his orange beanie on the dirty floor, and between sips from a chipped shot glass, he regaled my buddy and me with fantastic stories of diving adventures and legends late into the night.

In the days that followed, I realized that something miraculous had happened. My predive jitters had vanished. My hands no longer shook while gearing up. I was no longer "feeding the fish" between dives. I felt the most peaceful I had ever been while diving. Peace beyond my understanding. I believed God would take care of me, and so I felt entirely safe, even while doing some of the most dangerous things I had ever done. I wasn't unafraid, but the fear I felt held no power. It was helpful to remember what God had already done, all the times he had been trustworthy.

> Can anything ever separate us from Christ's love? Does it mean he no longer loves us if we have trouble or calamity, or are persecuted, or hungry, or destitute, or in danger, or threatened with death? . . . I am convinced that nothing can ever separate us from God's love. Neither death nor life, neither angels nor demons, neither our fears for today nor our worries about tomorrow—not even the powers of hell can separate us from God's love.
>
> ROMANS 8:35, 38

Christ is the perfect dive buddy, the greatest rescuer of all, the one who supplies us with air. He's the one who knows whether we are okay without asking, who is willing to go however deep it takes to get to us, who holds tightly on to us, and who always knows how to guide us back to the surface and into the light.

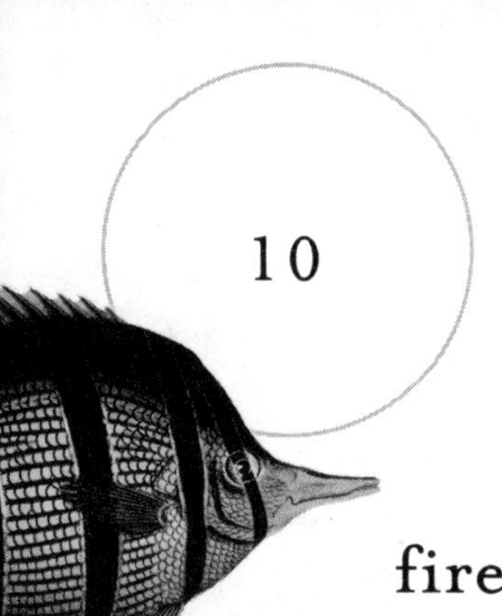

10

fireworms and butterflies

Bleary-eyed as the first rays of dawn combed through the trees, I was startled by a jolting movement and rustle of leaves on the path before me. It had been an otherwise normal morning at the university field station. I had arisen to my 5 a.m. alarm and was making my way from the cabin where I lived to the kitchen, where I would don an apron to prepare breakfast for the students, professors, and researchers who were still sleeping. Scrambled eggs, bacon, porridge, fruit salad, and hand-stuffed eclairs were on the menu and my mind. So when the trail path before me seemed to move, snort, and kick up leaves, I nearly jumped out of my skin.

Then I saw her—a deer. A small thing, as all the island deer were, but this doe was only a year or two old. She was lying on her side, a cluster of crisp leaves pressed firmly into her back fur from the slight rolling motion with which she attempted to stand. The poor thing could not. With eyes wide and panicking at the sight of me, she grunted and briefly pumped her legs before abruptly stopping, as if exhausted.

"It's all right. I'm not going to hurt you." I slowed my movement and gingerly tiptoed off the trail, making a wide berth around so as not to frighten her further, but close enough to study her.

"Something is very wrong," I quietly told the doe. There were no signs of injury—no visible wounds and no broken legs. She looked skinny, but it was normal for the island deer to be thin, as their population was so large and the amount of reachable vegetation limited. The doe's head had plopped back down on the trail, and she was breathing laboriously. It was clear that she could not get up. I hypothesized that some sort of brain injury had taken place, perhaps a seizure or aneurysm. Maybe given a little time, she would recover. I knew no harm would befall her on the trail, so I continued on my way.

While hurriedly stuffing custard filling into eclairs before students arrived, I noticed one of my professors, Dr. Eric Long. He had set up camp at one of the dining room tables, casually sipping coffee and typing away at research publication edits on his laptop. Dr. Long was known for using memes and GIFs in his PowerPoints, for being a Lucky Charms cereal fanatic, for inviting students over to dinner with his wife and kids, and for his Sunday morning "nature sermons" at the field station. He had endured (and even seemed to enjoy) my endless questions about the connection between faith and science and subsequently had introduced me to the minds and legacies of the academics John H. Walton and Steven Bouma-Prediger, as well as Saint Francis of Assisi. Besides being my academic advisor and a good friend, Dr. Long was a wildlife ecologist. And I knew he was editing a research paper about the wild deer population on Blakely Island. If anyone could help a hurting deer, he could.

Just before breakfast began, I led Dr. Long up the trail to the doe. For a long moment, we stood at a distance, watching her quietly struggle. Slowly, Dr. Long took off his windbreaker and approached the doe, laying the jacket over her head. The darkness had a calming effect; her agitated movement ceased. In one swift motion, Dr. Long grasped the doe's two front legs with one hand, her two back legs with the other, and lifted her off the ground. Quickly, he walked to the open back of a nearby pickup truck and carefully laid her inside. Looking toward me, he quietly said, "I'll take care of things and let you know how it goes." I knew what that meant. But I also knew that if I wanted to help a deer, certainly Dr. Long desired to do so as well. For as well as I loved nature, he loved it better.

Dr. Long was absent from breakfast. Later that morning as I finished washing dishes, he approached me. "Come outside when you can. You'll want to see this." When I did, he was leaning against the fence, holding a jawbone scrubbed clean and freshly white. Smiling slightly, he tapped the jawbone into his hand with the kind of quiet excitement born from solving a mystery and experiencing a new facet of nature up close. Then he told me the story.

The doe had been suffering and wasn't going to get better. After loading her in the pickup, he and the field station's maintenance caretaker had driven the truck to the south end of the island, where nobody would hear the gunshot. Still blindfolded, she died peacefully and instantly. Being the curious, exploratory type who loves nature and desires to understand it, Dr. Long had been determined to discover what had ailed the doe. He did an autopsy, dissecting the body. And that's how he ended up with the jawbone, which he now handed to me to examine.

One side of the jaw was exactly as you'd expect it to be, clean lines with a curve, teeth intact. The other side was surprising. It was deformed, having a strange scoop-like growth, making it impossible for teeth to properly develop. When Dr. Long had pulled out the jawbone, a massive ball of vegetation had come with it, settled into the scoop-like deformity.

The deformity had made it impossible for the doe to chew on one side of her mouth. As her jaw developed over time, the deformity worsened. When she tried to swallow mouthfuls of vegetation, the food got stuck inside her mouth on the one side, creating a giant accumulation that must have been horribly uncomfortable. Eventually, the vegetative mass collecting in her cheek at the back of her throat became so large that she was unable to swallow food, except for minute amounts. She became so weak that she fell on the trail and could not get up. The poor deer had a jaw deformity that caused her to starve.

Throughout the years, that deer's story has stuck with me. I think about Dr. Long's gentleness and mercy toward the doe, and I think about how she must have struggled, even without knowing it, living her whole life unable to entirely swallow food. To be able to taste but never experience the full nourishment of a meal—surely, that deer had suffered.

As an ecologist, I have wrestled against the suffering I witness in nature. When something goes wrong, I long to correct it, to intervene. We did this for the deer, and it was a clear choice.

But in other situations, the decision of whether to intervene may be muddled and complicated. Sometimes I can recognize that by stepping in, I may only make things worse. Nature has a way of healing itself, so if we do intervene, we ought to remember that "to keep every cog and wheel is the first precaution of intelligent tinkering."[1] Ecological studies have revealed that intervention frequently yields mixed results. Sometimes human intervention saves the day; other times by trying to help, we create an even worse situation. But in the moment of intervention, it can be hard if not impossible to tell what the outcome will be. We make the best decisions we can with the information we have available, and then we work toward and hope for the best.

I thought about this a lot during intervention dives for treating SCTLD-infected corals. Admittedly, antibiotic treatment was the best and most effective intervention strategy available at the time, but we didn't know what the outcome of putting thousands of grams of amoxicillin into the ocean would be. We still don't (at least, not yet).[2] But nobody wanted to sit around while corals died. Park management decided it would be more regrettable to do nothing and watch reefs perish than to do something and potentially make things worse. It's in the clarity of hindsight that we opinionated critics can scornfully note that coral reefs continue to suffer and that if in twenty years we wind up battling a slew of antibiotic-resistant marine bacteria, then it will serve us right. We just created a bonus problem. Hopefully, nothing of that sort will happen. But if it does, shame on us if we act surprised.

Another reason I seek to address the suffering I witness in nature is because of my theology. Suffering grieves the heart of God; therefore, it grieves my own. This may not make immediate sense, particularly if someone does not personally know God, so let me back up and explain what I mean.

A fellow scientist and one of my dear friends once asked me, "How can you believe in a God who created a world with so much evil and suffering in it? I would rather choose to believe that there isn't a God at all than admit there might be one who promotes evil." I recognize the courage of asking that question. I believe we have all asked it, or at least some version of it. In light of our current world and personal experiences, it is not only a valid and excellent question but also a tremendously important one.

What my friend wanted to know—what we all want to know—is really another question in disguise: "Is God good or evil?" For if God is evil, my friend is right. God would not be worth believing in, and certainly not worth following. But if God is good, why is the world the way it is? How does that inform our understanding of God's character?

To understand whether God is good or evil amidst a world that contains undeniable evil, we need to start at the beginning. It was humankind's choice to sin against God, which introduced suffering to creation.[3] That was not God's desire or his choice. However, God allowed humans to make this decision, presenting an option absent of his presence, a void separate of himself, an antithesis that all things opposed to him could move into.[4]

The presentation of this choice does not mean that God is evil; but rather, it demonstrates his love. Real love does not force or coerce, and so he gave us the ability to decide things for ourselves. Desiring to be in genuine, mutual relationship with his creation, God gave humans free will to follow him or walk away. In the Garden, Adam and Eve essentially chose to turn from him. "This means that the evil we see in the world is permitted by God's sovereign nature because He created people with libertarian freedom," says Phylicia Masonheimer. "God did not desire the world to have such evil, but in order for humanity to have the freedom of choice and relationship, they have to have the freedom to choose to move away from Him."[5] This choice continues to take place across time. In the same way that Adam and Eve could choose to reject God, people today have the choice to either continue rejecting God or be reconciled to him. God has known from the beginning how things would turn out, but because of his love, he still allows humans to choose for themselves whether to be in relationship with him or not.

From here, we need to realize two things: what love actually is and who is at the center of things. As C. S. Lewis reasoned, "The problem of reconciling human suffering with the existence of a God who loves, is only insoluble so long as we attach a trivial meaning to the word 'love,' and look on things as if man were the centre of them."[6] In other words, when God defines love (instead of the other way around, with "love" defining God), our perspective shifts. Our interpretation of "God being loving" is no longer defined by the flaky or unsteady versions of love we so often experience in life. Instead, love becomes something much more solid and graciously unmoving. The love of God is patient, kind, and enduring.[7] And so if God's love is not trivial but instead wholly sacrificial, desiring to "forgive all infirmities and love still in spite of them; but . . . cannot cease to will their removal,"[8] then we can finally start to see how a good God could exist over a cruel and suffering world, even as he continually grieves over the evil it contains. We can also see why God longs for us to leave our wrongdoing behind and be fully reconciled to himself. He loves us despite our evil, but still he wants it gone; for us to flourish, it must be gone.

God has not abandoned us. He could have done so, but because of who he is, he provided a way to bring us into the beautiful, sinless, eternal lives that we were originally intended to experience.[9]

God (being all-knowing) knew beforehand what our choice would be (that we would choose to reject him), but in his perfect love for us, he planned and provided the ultimate cure to the disease our choice created. He provided salvation from our sin, giving us a second chance at relationship with him. God loved us so much that he created a way—through Jesus—to both forgive our infirmities and remove them "as far from us as the east is from the west" (Psalm 103:12). And he did this because he loves the whole world. "For this is how God loved the world: He gave his one and only Son, so that everyone who believes in him will not perish but have eternal life" (John 3:16). Jesus willingly died in our place; his sacrifice for our sins overcomes the corruption and suffering in this world.

Our Creator performed this phenomenal cosmic intervention across spiritual and physical landscapes so that we can be redeemed from evil, restored to his originally good design, and reconciled to relationship with him. Because of his intervention, all of creation will one day be restored to its original good design as well. This rescue rewrites the meaning of our lives on earth and the endings to our stories in spiritual and physical ways, on a scale of eternity. Receiving salvation from Jesus spiritually restores us in life. In death, his salvation will physically restore us, just as Jesus was restored when he resurrected.

Until then, we remain in a broken world where we can experience the consequences of sin and death, even as we are freed from their grip and able to begin the journey of walking in intimate, personal relationship with God. With this perspective of the cosmic story, we can finally see love as it applies to God's character. Maybe now we can begin to wrap our minds around "why the world is the way it is," being, as my friend said, a place with so much evil and suffering.

God is perfect in everything he does[10] and perfectly loving.[11] There is no darkness in God,[12] and evil was not created by him. When he created the world, everything was "very good!" (Genesis 1:31). And because of him, someday all things will be made new, without any shadow of evil in them.

The world we see now is stretched between the sinful choice of humans and the future restoration God is bringing about. Our place in it is confusing with stark juxtapositions of life and death—and so much suffering. It reminds me of an experience I will never forget on a SCTLD intervention dive. I couldn't make sense of it then, and even now as I mentally work through the theology and biology of the situation, I struggle to feel satisfied.

I was cruising through the water, looking for infected corals to smother with amoxicillin paste, when I encountered a bizarre sight. The mazelike ridges of a mustardy yellow and green *Pseudodiploria strigosa* colony were covered with a thick strand of fluffy, feathery stuff. In fact, the fuzzy thing was moving like a caterpillar over the colony's disease margin, where the coral was actively dying. I could identify the disease's edge since there was a clear delineation between colorful live tissue and bright white skeleton. The fuzzy caterpillar-like creature was several inches thick and about the length of my forearm. It had vivid red feathers tipped with pale pink and white pluming out from joint-like segments. I poked its hind end with the eraser of my pencil, and the creature retracted its body and coiled away. By now I recognized it as *Hermodice carunculata*. This bearded fireworm is known for the venomous neurotoxin it can release from its fuzzy bristles and for its never-ending hunger for a tasty coral snack. The worm was not alone. Several of its fireworm friends were nearby, all feeding on the coral's disintegrating tissue at the disease margin, where the coral's body conveniently peeled up in little flaps off the skeleton. It was easy pickings.

As the fat and happy fireworms cruised their way across the coral, munching mouthfuls of diseased tissue as they went, I wondered if and how they would be affected by the amoxicillin I was about to apply to the sick coral. I scooted the fireworms off the colony with the end of my pencil, careful not to touch their venomous bristles, and marveled at the strangeness of the ecological niche they occupied. The coral's swift death at the hands of SCTLD provided a quick and convenient meal for them. What was bad for corals was wonderful for fireworms. And not only the fireworms, but butterfly fish too.

Hundreds of times I had seen beautiful foureye and spotfin butterflies with delicate lips plucking polyps from coral disease margins. I hadn't thought much of it, since these fish tend to eat whatever is conveniently available, but the thought settled over me in light of the fireworms. The corals were food for these creatures, and even the disease itself. Although we had not yet determined the exact microbial culprit of SCTLD, we knew that it was likely a suite of bacteria and viruses acting in tandem. The corals' suffering and demise was these microbes' success.

My mind began to spin. Having a bacterial component, the disease was a living thing, too, just like fireworms and butterflies. What gave corals any more right to live and thrive than these microbes? Was it wrong for fireworms and butterflies to feed on coral flesh? Was I wrong to value one organism over another simply because it was more beneficial to people? Surely all these creatures were created wonderfully, but then why was the successful function of one so detrimental to the other? It is a world of death and life, systems and cycling, give and take. All these seemingly contradictory juxtapositions. I wrestled with these as I cautiously poked fireworms, gently swatted at butterflies, and pasted antibiotic over the top of diseased corals.

God provides food for every creature and yet loves all things, including fireworms, butterflies, corals, and the deer who collapsed from starvation. While this may not wholly satisfy our limited understanding, we know this is true because the Bible confirms it: "Give thanks to the Lord, for he is good! . . . He gives food to every living thing. His faithful love endures forever" (Psalm 136:1, 25). This is yet another confirmation that God is good, even the very definition of good, just as he is the definition of love. Elsewhere the Bible says, "He gives food to the wild animals and feeds the young ravens when they cry" (Psalm 147:9). God provides food for every creature, including carnivores and scavengers that eat other animals—further evidence that God permits rather than promotes the suffering we see in the world.

The deer that died of starvation was a casualty of sin in the world. Similarly, perhaps the habits of carnivores and scavengers are the result of sin, from humankind's free will choice in the Garden of Eden. Or perhaps carnivores and scavengers are actually performing their original design, since there could be another piece of this mystery that we haven't made sense of yet. In a rhetorical question spoken by God, humans are asked, "Can you stalk prey for a lioness and satisfy the young lions' appetites as they lie in their dens or crouch in the thicket? Who provides food for the ravens when their young cry out to God and wander about in hunger?" (Job 38:39-41). Carnivores like lions "roar for their prey, stalking the food provided by God" (Psalm 104:21). These creatures know full well who their true provider is. From wild donkeys to flying storks, from rock badgers to the Leviathan, all creatures look to God to give them what they need at the proper time.[13] God provides for and loves all things, even when he does so in ways we don't expect or understand.

My idea of God is not an entirely correct idea. I am a willful human with a sinful heart in a broken world, trying to comprehend a perfect, righteous, loving Creator. I am trying to make sense of his creation and how it points to his character. I find a deer starving and recognize that things were never meant to be this way. I watch predatory worms and fish eating an animal I love and wrestle with deeper questions. Try as I might, my view of God and the world he created will need adjustment along the journey of our relationship, as I come to know him more. "My idea of God is not a divine idea," says C. S. Lewis. "It has to be shattered time after time. He shatters it Himself." Like Lewis, I rest in this truth because what I truly want is "not my idea of God, but God."[14]

Great is the Lord and mighty in power; his understanding has no limit. He provides food for the corals and creatures of the sea, opening his hand and satisfying them with good things. He takes their breath away and they return to dust; he breathes and they are created, renewing the face of the earth. The fireworms and butterflies call for their prey and seek their food from God.

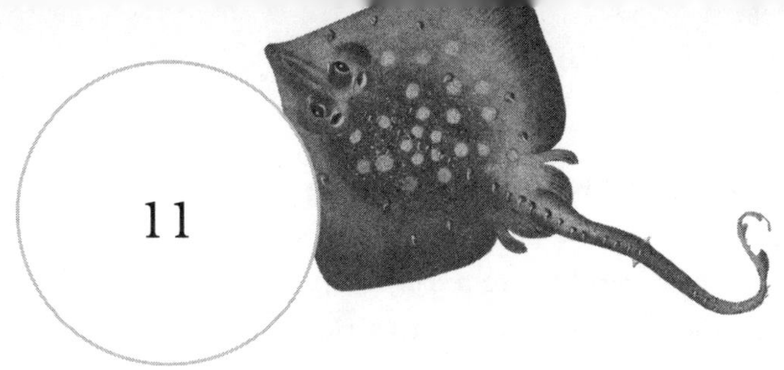

11

bungling the bailout

My toes curled over the boat's edge, water chaotically splashing through the dive door up to my knees. I could feel my heart rate steadily increase as my fingers twitched anxiously against the heavy weight of dive gear. I was heading into the last, hardest, and most notoriously dreaded portion of our annual professional dive skills exam.

Why does a professional diver need to take an annual exam? The scuba industry is a tad bit peculiar. Anyone can go diving if they have access to the right equipment and a body of water. But just because it's possible, doesn't mean it's recommended. Getting proper training could save the diver's life. And this training is what constitutes a dive certification, which is comparable to a license.

Scuba certifications operate in levels: Students begin as open-water divers, able to dive to sixty feet. If divers continue to train, they'll gain an advanced

certification, able to dive to 100 feet and perform skills like underwater navigation and shipwreck penetration. Next is a rescue certification, teaching divers how to rescue themselves and others in dangerous situations. (This course in particular is considered a must-have among professionals.) Certifications top out at a divemaster certification, where divers become qualified to teach others how to dive. But the learning never really ends, as there are a slew of side courses, programs, and schools that teach different skills: technical diving, maintenance diving, nitrox diving, and underwater photography, to name a few.

Once a diver is certified, they are certified for life. But while they might be free to dive whenever they want, it may not be safe to do so. Scuba is all about immersing ourselves in an unbreathable environment, so regularly brushing up on the skills that keep us alive is something all good divers practice. Not to mention, one of the most important aspects of dive safety is fitness. Physical, mental, and emotional fitness is one of the major distinguishing factors between recreational diving (the kind that people do "for fun") and professional diving (done "for work"). In order to determine if divers remain fit, they need to be regularly tested. Among professional divers, dive exams are frequently integrated into the job.

This was the case for me as I stood at the boat's edge holding all the scuba gear in my arms, save the wetsuit and dive boots I was wearing. I wore no mask on my face to help me see, no fins on my feet to help me swim, no weight belt around my waist to help me sink. My thirty-five-pound cylinder, the only air source once underwater, was turned off, and I clutched my regulator in my hand.

I knew what would soon be expected of me. I would exhale, pushing excess air out of my lungs to become as negatively buoyant as possible, holding tightly to my gear so as not to lose any life-sustaining equipment, and then bravely jump into the water. In a James Bond–style maneuver, I would use the streamlined momentum of my leap to swim to the sandy bottom of the seafloor, twenty-something feet below. Quickly turning on my cylinder, I would insert the regulator's mouthpiece and breathe. Then I would begin to systematically don each piece of equipment: mask, dive belt, fins, BCD. Once properly suited up and adjusted, I would calmly ascend back to the surface to be greeted by cheering teammates on the boat.

Some things could go wrong, of course. When jumping off the boat, I could potentially drop a piece of equipment and have to start over. Or I might not be able to sink and swim to the bottom; I might strain to violently kick my way downward for several minutes to no avail, my lungs and throat burning from lack of air. In that scenario, I could become so exhausted I'd have to resurface, only to slam my head against the side of the boat. Or worse, I might bash my head against the stationary propeller of the boat. If that happened, there could be blood.

This exercise was called the bailout. And every person to ever achieve this physical feat has wondered why it was necessary. Were we practicing a getaway maneuver in the extremely unlikely case our boat would ever capsize or explode? Was this some bit of reapplied training from our burly dive officer's days in the Navy? Or was the whole thing a cruel joke drafted by the head of federal dive leadership as a means of making field divers pay the price for the desk-chained bureaucrat who missed the old, salty days of glory? When our dive leadership team was asked, they'd answer, "It's probably not something you'll ever need to do in real life, but it's an advanced skill that showcases your ability as a diver. It's been on the test for years." Whatever the reason, either as an extreme safety procedure or comedic story potential for higher-ups, the reality of the bailout was no joke.

Some divers were skilled. Some were lucky. The first year I performed the bailout, I was Bond—James Bond—and finished the exercise in a single try. The story became a sort of soft legend among our dive team and earned me significant respect. But when the year lapsed and we were retested, my status as bailout whiz promptly switched to bailout clown. I tried six times unsuccessfully, hit the propeller, and bled.

Sitting on the edge of the boat with a towel pressed to my scalp, sinuses burning from salt water, legs rubbery from exhaustion, I felt embarrassed and afraid. I had to complete the bailout because it was a requirement for my job, so I would have to try again the following day. My anxious mind turned to a defensive, warped sort of logic: *This is just a ridiculous talent show to humiliate people with test anxiety. We'll never use this skill when we're actually diving. Why do I need to do this?* The thoughts felt so bitterly briny as to provoke a gag reflex. But honestly, I knew the deeper truth of why. I hadn't made it to my second bailout without already having encountered much harder things along the journey.

In tenth-grade biology class, I was a typical awkward fifteen-year-old, perched on a wobbly stool and wearing a white lab coat while my teacher, Mr. McCormack, waxed poetic about the chemical components of glucose. It had not escaped my attention that Mr. McCormack's tuft of red hair looked remarkably similar to that of Beaker, my favorite character from the Muppets. Mr. McCormack was infamous for his tangents, which I usually found more interesting than the topics we were supposed to be learning. When we were meant to be hearing a lecture on glucose, we somehow wound up discussing the Mariana Trench, hydrothermal vents, and deep-sea tube worms. I sat, trying not to wiggle conspicuously on the uncomfortable stool, and hung on Mr. McCormack's every tangential word.

I didn't always like school, but I loved learning. To my way of thinking, teachers were the guardians of the knowledge I craved, like doorkeeper guides to a magical land I longed to explore. I believed that if a student approached their teacher with humility and curiosity, the teacher would undoubtedly usher them in and happily mentor them in everything they knew. But if the student acted like an entitled brat or incurable fool, they would have to fend for themselves without help from the teacher. Because of this belief, I was a driven, hardworking student with an ultrahigh desire to please teachers. A super nerd and teacher's pet rolled into one, I was the kid who was sincerely excited to write essays and read the books we were assigned. I had a natural talent for literature, language, and art, and because I never shared my grades with friends, it was assumed by most everyone (besides my parents and teachers who knew better) that I was a top-performing student. The other kids didn't know I was flunking tests. And besides the plague of paralyzing test anxiety that consistently dragged my GPA, there was another major problem.

I was bad at science. I loved it, but I barely understood one lick of it. The sight of autumn leaves changing color, the smell of fresh rain, the feeling of sand on the shore . . . *these were more than miracles to be accepted at face value*, I thought. They were invitations to wonder, question, and learn. Science could help me do that, dazzling me with its ability to identify and explain the world that I so desperately wanted to know. But science was my worst subject, and even when giving it my all, I would just barely squeak out a passing grade. Simply put, I wanted to be smart, but I couldn't figure out how.

On the second day of tenth-grade biology, Mr. McCormack went off on a tangent about graduate school research he had done back in the day. He and a team of scuba divers had lived in a submerged underwater tank to experiment with the effects of ultraviolet radiation (levels of sunlight) on corals. They found that moderate amounts of light helped the corals grow better than low amounts but that high amounts caused the corals to bleach. Mr. McCormack also told us about the symbiotic relationship between coral and a special kind of algae, called zooxanthellae, that lives inside coral tissues. The corals give zooxanthellae a home, protection, and nutrients, while zooxanthellae gives corals food from its photosynthesis process and color from its own pigmented body. But coral bleaching completely unravels the symbiotic relationship between algae and coral. When corals become stressed (usually from high temperatures), they expel the zooxanthellae from their bodies, causing all their color to leave with the algae. This turns the corals white, hence the term *bleaching*.[1] And although corals are still alive when they bleach, they are far more likely to die if the source of stress continues. My mind was completely blown as Mr. McCormack's in-class tangent became a detailed lecture on the biochemical workings and ecological implications of coral bleaching.

From this moment on, two things were clear to my fifteen-year-old brain: Coral reefs were the absolute coolest, and they were in trouble. I thought about the wall-sized map of the world in the social studies classroom, with a large, pinky-orange swath of color stretching down the east coast of Australia and similar speckles of pink in places around the Caribbean and Florida. Coral reefs were so big and important that they were included on maps. I wanted to see them and help them.

When my mom picked me up from school that day, the first thing I told her about was coral bleaching. I talked so fast that my words tripped over themselves, and suddenly tears were slowly running down my face.

"I don't know why I'm crying," I said, sniffling. "Corals are just so amazing, and I'm really excited!"

My mom's sweet reply was something I will never forget: "Wow, you really love this. Maybe you should remember that."

And so I did.

I trace my career path back to the second day of tenth-grade biology class and that moment of crying in the car. It was the vocational call on my life, which God confirmed consistently in a variety of ways through my high school education and gap year at an international Bible college. When I entered my undergraduate program at age twenty, I had a crystal clear vision of what I wanted to do and zero idea of how to get there. Passion was my fuel, and I figured someday it could take me all the way to the moon—or at least to the Great Barrier Reef. But it wasn't long before I realized I would need more to get there.

My college advisor was a tall, long-haired man with a friendly face who, after listening to my excited coral-related exultations, suggested with a laugh that I major in ecology. (This man was Dr. Long, the professor who would play an important role in shaping my understanding of how faith and science intertwine, which was a topic I wrestled with during my first years of college.) I took all the basic required ecology courses: biology, chemistry, calculus. But with a full course load of subjects I was naturally bad at, it didn't take me long to realize I was in over my head.

My grades rattled me to my core. I was not on track to pass biology, so I went to my biology professor's office to get help, multiple times. While clearly gifted in his field, he had none of the easygoing, encouraging manner of Dr. Long. Even so, he did give me study skill ideas and extra practice problems. I worked harder and continued to struggle, right up until the final exam. The moment of truth.

Right after the final, I went straight to his office, trying not to cry as I waited with my back against the wall. I knew deep down that there was no way I had passed the test. My old, undefeated dragon of test anxiety had reared its ugly head again and defeated me. How was I supposed to be a marine biologist when I couldn't even pass a general biology course?

I viewed my professor as the guardian of knowledge, and I wanted his advice because I admired him. I had excitedly told him about my dream of being a marine biologist because I longed for his mentorship. When he finally arrived at his office, he took one quick look at me, turned his face away, and clutched his briefcase a little tighter. I quickly wiped my nose, stood, and said hello, but he greeted me with silence while unlocking the door and stepping inside. I wasn't sure whether to follow.

After several long moments, he told me to come in. I shut the door, sat down, and unsuccessfully tried to suppress tears while saying that I knew I hadn't passed. I had tried my hardest and wanted to know what to do. He took a red pen and stack of papers from his briefcase, thumbed through a pile of exams, and made a series of red dashes on one of the test packets. With my exam score figured into my final grade, I was 0.5 percent away from passing the class. So close, but not good enough.

I thought this professor was brilliant, and from the beginning, I had sincerely longed to win him over. I recognized now it had been to no avail. I felt false pity in his stoic gaze, and some part of me realized that he must be an altogether different species from the sort of teachers Mr. McCormack and Dr. Long were. While I was far from a flawless student, I had truly tried my best, and the real problem was that my best wasn't good enough. Now here I was, begging for his help. "What should I do?"

His simple reply shattered me: "You should switch majors. If I were you, I would seriously consider changing career paths."

My professor's words hurt. Then I got really mad, the sort of mad where you light a bonfire so hot under your own bum that you flee from the flames. I fled from that professor's advice because if God had called me to be a marine biologist, then who would prevent me from becoming one? Certainly I would not let myself get in the way of God's will for my life. Why would I let anyone else get in the way either?

As God gently uprooted the bitterness that was trying to take hold in my heart, he planted a version of stick-to-itiveness I had never before possessed. My parents and friends called it tenacity. I became even more firmly rooted in my ecology major, retaking that biology class and passing with a near-perfect grade. My old biology professor watched from the shadows.

Granted, during the second semester, I bombed my second-level chemistry class. But I was wiser and more tenacious now, and my chemistry professor was remarkably kind. I retook the class and aced it, knowing that if God had brought me this far in the face of adversity, he would take me the rest of the way, however far that was.

He did. By the time I finished undergrad, I was a study fiend. Some of my professors had become my closest friends. And I had learned to live in harmony with test anxiety. Even outside the biology department, I was known for the goofy pep talks I would give myself before exams, talking out loud to myself in the restroom nearest to the testing classroom. Sometimes my female classmates would gather with me in front of the bathroom mirrors to listen, saying it inspired them too. With hands on my hips and the stance of Wonder Woman, I would make fierce eye contact with my own reflection and state with the seriousness of a general directing her troops before battle, "You have studied hard. You know the information. You serve a limitless God who has called you to do this, and he can enable you. He already has. Because of what he has called you to, you will do it. You want this more than anyone."

The things that limit us—even things as basic as water, air, light, and knowledge—are not limitations to God. As the creator of everything, God is beyond everything. He cannot fail as we do. He does not get confused or lost. He will not drown or suffocate. He does not fail any test. But he can enable us through his ability, infinitely beyond our own, so that we can do far greater things than the wildest stunt performed by scuba divers or James Bond.

I can look back on the struggles of undergrad and see how God was using them to prepare me. Taking exams amidst test anxiety helped me find ways to work with fear instead of against it, while developing even more efficient study habits. As a result, when I studied at an international graduate school where the exam format changed from multiple choice to longhand essays, I became academically successful in a way that didn't make sense to my STEM-talented peers. I finally found the niche where my tenaciously cultivated scientific knowledge met my natural talents of writing and language. Finally, I was good at science!

I never truly overcame test anxiety, but learning to live alongside it prepared my mind and body for the stress that accompanies technical scuba diving. When you're used to mind games and adrenaline rushes on land, you're better equipped to handle them in a calm and safe manner underwater. I also learned to screen carefully for trustworthy mentors and not be rattled by naysayers.

All these were skills I needed later in graduate school, my first job, and beyond. My journey of becoming a marine biologist is best summed up by the words of A. W. Tozer: "How completely satisfying to turn from our limitations to a God who has none."[2]

Now years later, during the boat ride to the site where I'd retake my bailout exam, the inner tension of my test anxiety slowly built. I prepped my gear in determined silence, feeling the tightness of the scab on my scalp from my first failed attempt. Then I stood facing the horizon, hands on my hips, mouthing the words of my battle cry: *I serve a limitless God who has called me and will enable me. Because of what he has called me to, I can do it.*[3]

I grabbed my gear, expelled all the air from my lungs, jumped off the boat, and swam to the bottom like a fish. Underwater, I donned my gear calmly and swiftly, then ascended perfectly and was congratulated for completing the exam.

I've never needed to do a real-life bailout. There has never been any instance in which I couldn't suit up on board. There was never any call for me to ditch and redon every piece of my dive gear underwater. There was never any sudden threat of a boat exploding, leading to a hasty evacuation. But the bailout did do something important for me. God used it to get me ready for future challenges.

The bailout prepared me for the time that I accidentally forgot to turn on my air supply before entering the water. When I tried to take a breath twenty feet below the surface and found there was no air to be had, I didn't panic. Instead, I calmly reached back to turn the valve and started the flow of air. It was no big deal, and I wasn't afraid because I knew I had done harder things. The bailout prepared me for the time my dive mask was knocked off my face and fell under a rocky ledge. I carefully swam down and retrieved the mask, put it back on, and exhaled through my nose to purge it of water. Granted, I lost one of my contacts and got some water up my nose in the process, but the discomfort wasn't a new sensation. I had already experienced it. And the bailout prepared me for the time that a less-experienced dive buddy ascended without checking the surface to see where the moored boat was. He was going to hit his head on its stationary propeller. But I saw the danger, remembered the pain of my scalp, and signaled him out of the way in time. The previous hurt I'd experienced made me look out for danger. Because of that, I knew what to do.

Sometimes God calls us to uncomfortable things that create physical, mental, emotional, and spiritual discomfort. These things may push us forward, require a scary jump, make it difficult to breathe, or take us to the depths. We might not always understand why. We might struggle with anxiety or doubt. People we look up to might encourage us to quit. But we can be assured that if God has called us to something, he goes with us. He jumps off the boat beside us. He bashes his head into the propeller with us. He goes to the depths of his ocean with us. He will never leave nor forsake us.[4]

"Be sure of this: I am with you always, even to the end of the age" (Matthew 28:20). These were Jesus' parting words to his disciples as he gave the great commission. Interestingly, it's noted that although the disciples saw and worshiped him, having already experienced the power and promise of his resurrection, some were still struggling with doubt.[5] But throughout the course of Jesus' ministry, he had been preparing his disciples for this moment when he would send them into all nations. Even to those disciples who doubted, Jesus reminded them of himself. He comforted and encouraged them, not with a list of the skills, credentials, abilities, or qualities they possessed, but with the truth of himself, as the one with "all authority in heaven and on earth" (verse 18). Alongside what was likely a scary, tremor-inducing send-off into "all the nations" (verse 19) to make disciples, Jesus promised that he'd be with them always.

With the same being true for us today, we might rhetorically ask God,

> Where can I go from your Spirit?
> Where can I flee from your presence?
> If I go up to the heavens, you are there;
> if I make my bed in the depths, you are there.
> If I rise on the wings of the dawn,
> if I settle on the far side of the sea,
> even there your hand will guide me,
> your right hand will hold me fast.
> If I say, "Surely the darkness will hide me
> and the light become night around me,"
> even the darkness will not be dark to you;
> the night will shine like the day,
> for darkness is as light to you.
>
> PSALM 139:7-12, NIV

My story is exactly that. When I went to the depths of the sea, rose with the dawn, and lived abroad, God was there with me. Despite all my limitations, he has proven himself limitless.

I still remember the longing in my heart as I stood in my high school social studies classroom and stared at the map that illustrated the locations of global reefs. I never would have made it to any of those places if God had not taken me there. He led me to the pinky-orange swath east of Australia. I saw the Great Barrier Reef often during graduate school in Queensland, driving boats in the Palm Island Group and snorkeling the Whitsundays. I visited some of the pink speckles around the Caribbean when I took a coral reef ecology course in Belize. I even dove the Great Blue Hole. And I came to intimately know the Florida Reef when I got my dream job as the coral biologist of Dry Tortugas National Park, which is where I met a friendly porcupinefish at the Windjammer, enjoyed exploring the shadows of the Maze, was nipped by a red grouper while sampling corals, and bungled my way through the bailout.

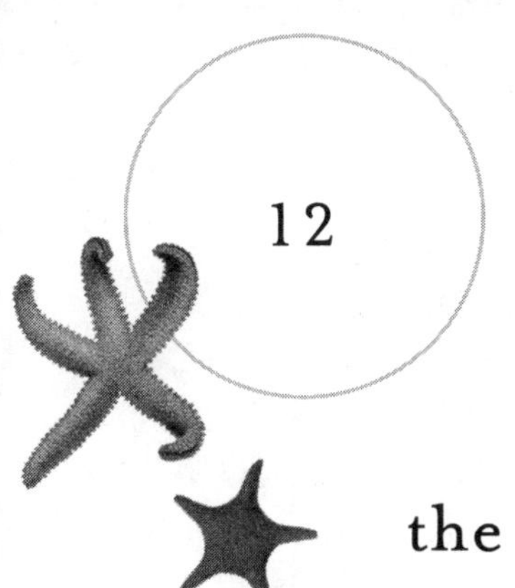

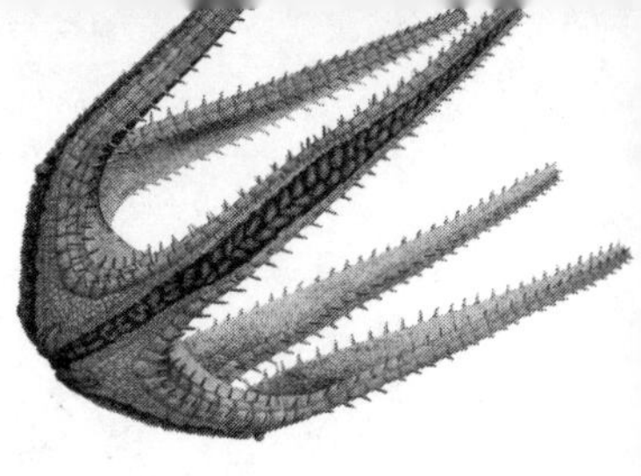

12

the betta hospital

In a blaze of glory, I graduated with my master's degree from an Australian university by the Great Barrier Reef. I finished in the top 5 percent of my class, leaving with enthusiastic commendations from my professors and recognition for my publicly presented thesis: "Calcification Chemistry Potentially Links Dimethylsulfoniopropionate (DMSP) Metabolism and Rapid Skeletal Growth in *Acropora tenuis*." (Yes, dimethylsulfoniopropionate. It is a real word for a special kind of sulfur molecule that corals break down in a very cool chemical reaction. This reaction produces the hard substance acrylate, which some corals can choose to incorporate into their skeletons, and the gas dimethyl sulfide, which they release back into the atmosphere.)[1] Needless to say, when I finished grad school, I was feeling pretty decent about my accomplishments. Then reality sank in.

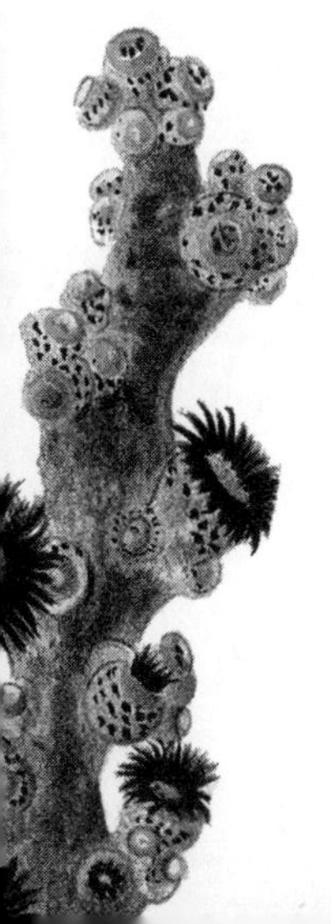

I was an American in Australia on an expiring student visa, so I needed to return to the United States. The remaining days of my sunshiny life Down Under became a long series of hard goodbyes—leaving my peers, my professors, my church, my home. On the bright side, I looked forward to being back with family. However, at about this same time I learned that my family at home was going through an extremely difficult period, a sort of heart-wrenching turmoil that extends beyond this particular story. The important part to know is that it was the hardest thing we'd ever gone through, and that for most of the journey, I didn't think we'd make it out still together.

On top of all that—the teary goodbyes and heartbroken home—I wasn't landing a job as a marine biologist. I honed my CV and my cover letters, I had them proofread by friends and professionals, and I cold-called for every opportunity. I had glowing recommendations galore, I became an expert at navigating online application portals and AI filters, and I was throwing job applications around like confetti. I was doing everything, all the right things, and I still could not get a single interview.

Weeks passed; months passed. I spent the days crying and healing with my family, and I spent the nights crouched over my computer until 3 a.m., sending out application after wretched application. The rejection letters flowed. Most were copied-and-pasted templates that read nearly identically. Most scratched my tired heart with a thousand paper cuts, but some were kind. "Keep trying; you're a good candidate and someone will pick you." But they hadn't picked me. Those rejections cut in a different sort of way. Worst of all were the no-replies and the open-ended maybes that never became yeses. Those left me straining to balance on a rapidly decaying thread of hope.

I felt unwanted, unhelpful, stuck, and miserable, anything but the heroine of one of those success-story movie plots. In those films, the story generally skips over the long, hard waiting period using a quick-paced montage of scenes backed by peppy music to demonstrate how hard the zero-to-hero character is working. Movies can take the darkest, stillest moments of life and run them together in a flashing series of distractions, cutting to the brilliant success. After all, we love to dwell on successes and not the horrible, stagnant yet emotional roller-coaster ride we take to get there. My life has not been like the movie montage version, a series of tiny, meaningless moments of working and waiting for success. Some of the greatest things happened during my montage, things I wasn't able to skip over, even though I wanted to.

I got a job at the local pet store. I figured it would get me out of the house, give me something to do besides hunch like a goblin over a computer in the middle of the night, and put some money in my dry bank account.

I wasn't in the mood to like my job. It seemed embarrassing. I had a master's degree, and here I was making just above minimum wage at a pet store, where all my bosses were less educated than I was and almost all my coworkers were high schoolers. And yet I loved it. My managers were kind, my coworkers were smart, and I was hungry to learn anything they had to teach me—which turned out to be quite a lot.

I was assigned to the aquatics department, and after a couple of months I was given free rein to do pretty much whatever I thought was best. My recent high school graduate supervisor was a genius aquarist and one of the most passionate hobbyists I've ever met. He taught me how to design and build tank systems, different techniques for cleaning and maintaining aquariums, and the taxonomic and common names of many freshwater and marine species. Thanks to him, I learned how to safely catch spiny, toxic fish without stressing them out, how to troubleshoot customers' tank problems and provide solutions, and numerous other skills. My frustrated restlessness was quieted by the bustle of my supervisor's mile-long to-do list for the department: unpacking new shipments of fish, feeding in the mornings and evenings, serving customers, cleaning tank systems, restocking merchandise, giving medical treatments to sick fish, and receiving betta shipments.

Bettas are tropical freshwater fish, native to southeast Asia and known for their flowing fins and tails as well as their brilliant colors. Males are highly territorial, so they must be kept in separate tanks; if put together, the fish fight to the death. Despite their aggression, bettas are a fan favorite because they are highly charismatic and relatively simple to care for. They're often bred for specific traits, such as extra-long fan-shaped tails, multicolor splotches, and striping patterns.

Each betta was shipped to us in exactly two tablespoons of water inside an equilateral triangle of a plastic bubble, about two inches from end to end. These triangle bubbles were loosely arranged in a cardboard box stuffed with crumpled foreign newspapers—sometimes K-pop ads, sometimes propaganda. I was horrified to discover that the fish traveled across the planet in a plastic bag containing less water than someone would drink in a normal sip. Sometimes the triangle bubbles would fold over on themselves during shipping, sealing fish off from the meager amount of water they had access to. Without water, their gills stuck together and they would suffocate. Some fish would arrive gasping, barely alive, their water brown with poop and decay.

Every two to three weeks, a new shipment arrived with the

morning deliveries. That gave me an entire shift to evaluate the health of the new fish. The survivors got bowls of dechlorinated fresh water and a big meal. The healthy ones swam around happily, excited to be out of their two tablespoons. These fish immediately went onto the shelf for sale. There was nothing I could do for the dead fish. Out of a shipment of fifty, anywhere from one to twenty-five could be dead. Several times an entire shipment came dead on arrival (DOA). The little bodies went into a plastic bag to be placed in the biohazard freezer, where they would eventually be disposed of properly.

As heartbreaking as the dead fish were, the sick bettas posed perhaps the greatest challenge to the department. Recovery was rare. We couldn't sell sick fish; if they weren't properly treated, they'd usually be dead within a day. Then they'd join their deceased buddies in that smelly freezer. But so many of them—anywhere from two to ten fish per shipment—were sick. Every sick fish needed to be treated and healed within two weeks, before the next shipment and more sick fish arrived. That wasn't possible. We didn't have the space for more than two or three extra quarantine tanks at the store, and we certainly didn't have the time to treat more than one or two individual fish with unique maladies during a shift.

As much as we cared for these creatures, trying to save them all just couldn't be a priority. Even so, it tormented me to see the condition that so many bettas arrived in. And every time a purple betta showed up DOA, I thought of my own beloved Vector.

In late high school, I got a vibrant purple betta. Vector was my roommate throughout most of college, greeting me when I got back to the dorm after classes, perching front and center in his tank while I did homework, snuggled up in his favorite plant. He became quite well-traveled, voyaging over 12,400 miles in his lifetime, riding shotgun in a pickle jar beside me on trips between university and my hometown. Vector became a mascot for the students on my floor. He also became a sort of student counselor, probably because listening to my dramas made him an expert for everyone else's. Some days, there'd be a knock at my dorm room door. "Oh, hi, I was wondering if I could see Vector. I'm trying to study for a test, but I'm really stressed out. I just want to watch him for a little while."

Seeing bettas like Vector who were sick made me feel helpless. The problem was bigger than me. Ordering from a different supplier would just bring another version of the same problem. I couldn't go to Asia and single-handedly demand they change shipping methods. And I couldn't challenge the pet industry without it reflecting badly on the good-hearted, well-intentioned people I worked alongside. But I had to do something; I just didn't know what yet.

For whatever reason, during this confusing stage of life, working at the pet store and staying up until the wee hours desperately applying for jobs I would never get, I couldn't get kintsugi off my brain. Kintsugi is a form of Japanese pottery where an intricate pot is created, becomes chipped or may be intentionally broken to pieces, and then is put back together with gold or another precious substance.[2] An artistic embodiment of restoration, the philosophy behind kintsugi is that places of brokenness are to be honored. The process of going from whole to broken to remade gives greater value to the object. A piece of kintsugi pottery would not be kintsugi if it had not been broken and then restored; it would just be regular pottery. It is the breaking and subsequent mending that makes it extraordinary art.

I thought about the pottery while applying for jobs; I thought about it while trying to tune out the sound of my family's voices down the hall; I thought about it on my lunch break. I was doing my best to trust Jesus, but everything hurt so much. I looked around and saw myself in a condition of brokenness. Breaking career path, breaking family, breaking heart. Kintsugi became the metaphor for the broken present in which I found myself. And slowly, as I worked long hours with the fish and had long conversations with my counselor and family, kintsugi also became the metaphor for the restored future I hoped to someday have, where God would somehow put the broken pieces back together, creating beauty in the process. I had zero idea how he would achieve that, so I kept applying for marine biology jobs, I reconnected with my family, and I kept showing up for my shift at the pet store.

Somewhere in the midst of being upset about a recent shipment of bettas, I noticed an issue in one of the saltwater tanks. One of our sand sifting sea stars hadn't moved since the day before. He was perched on the surface of the sandy bottom, motionless.

Given their name, it probably won't surprise you to learn that sand sifting sea stars sift sand. More specifically, they use the tiny, sticky tube feet on the undersides of their arms to move particles and clumps of sand. In a shifting, otherworldly crawl they are able to sift themselves down several inches through the sand to feed on buried and decaying particulates. So to see a sand sifting sea star sitting on the surface was quite concerning.

The sea star was about the size of my palm, so I scooped him out with my hand to take a closer look. The tip of one arm was missing. Where the arm had been, the edges of flesh were ragged, as if they had been chewed off by some tiny animal. Warily, I put the sea star in a small quarantine tank, free of decorations, sand, and other animals, and kept an eye on him.

In a period of twenty-four hours, over half his arm went missing. A fine bit of powderlike sludge could be seen in the tank next to him. The wound had turned necrotic. His arm was disintegrating. Although I was unsure of how he had acquired the injury, I knew one thing: the little sea star was doomed unless the necrosis was somehow stopped. Without intervention, the infected tissue would continue to unravel first his arm and then his whole body, killing the sea star in zombielike fashion.

This sounds pretty bleak unless you know that, like all echinoderms, sea stars possess a hydrostatic skeleton—a complex system of micro water pumps that helps them move. This lends sea stars a superpower ability: As long as a portion of their hydrostatic skeleton is intact, their limbs can completely regrow.[3] In fact, if a sea star is cut in half down the middle, both halves could regrow their missing half. One sea star could become two genetically identical individuals; the sea star would effectively be cloned.

The next day I got brave. And I told the sea star that he needed to be brave too. I put him in a petri dish and amputated the rest of his necrotic arm in a clean slice with my sanitized pocketknife. I dipped the wound in an iodine dilution and placed him back in a clean quarantine tank.

A few days later, the wound had closed up. I gave the patient some soft sand to play with, which seemed to make him quite happy. He burrowed down into it almost immediately. Besides his slower than normal pace and being down one arm, he seemed otherwise back to himself.

In a small way, the sea star reminded me of myself. At some point, we're all little sea stars, stuck on the sand with a necrosis we can't shake. We're in need of a loving amputation, for a Savior to cut away from us the infected parts that lead to death. We may lose parts that seem important to us, limbs we can't imagine living without. It will hurt. In the midst of pain, it may seem as if healing will never come. After all, how does a sea star know that his arm will grow back? I don't think he does. He is at the mercy of the caretaker who knows him, who sees both his doom and his salvation, and wields a loving pocketknife.

I realized that if God cuts something away from me, I want to be like that brave little sea star and sit quietly in the petri dish until my arm is gone. I want to obediently reside in the tank, even alone and lonely. Even as I have questions, "What are you doing?" "How can this be good?" "Why am I here?" "Don't you see that this hurts?" I want to rely on what I know to be true of God's character: that he is good, that he knows me, that he loves me. He has plans to prosper me and not to harm me, plans to give me hope and a future.[4]

I want to remember what the psalmist wrote:

> He heals the brokenhearted
>   and bandages their wounds.
> He counts the stars
>   and calls them all by name.
> How great is our Lord! His power is absolute!
>   His understanding is beyond comprehension!
>
> PSALM 147:3-5

Saving the sea star was just the beginning. As I watched him happily bury his remaining four arms in sand, an idea emerged that I couldn't seem to shake. I wanted to save every sick fish that came through the department. Especially the bettas who, up until this point, had definitely gotten the short end of the stick. Some of them wouldn't make it, but I would do my best to restore each of them to full health. If I couldn't save them, I would give them what comfort and mercy I could in death.

With my newfound, harebrained idea, I did some research and requested special permission from the store manager. Having approval from both work and my parents, I cleared everything off the dresser and side table in my bedroom. I sourced 6 one-gallon glass vases from my folks' garage and from a craft store with a coupon. I used my employee discount and part of a paycheck to buy the best-quality water conditioner, food, and treatments for every possible malady, and six soft-leaved live plants, one for each quarantine tank. While the plants were arguably not necessary, I knew they would significantly increase comfort for the fish in their otherwise empty vases.

My bedroom became a makeshift treatment center for bettas. On the next shipment day, I started bringing sick fish home, placing each into its own quarantine tank. On my own time, each fish received daily health checks and treatments, with the hope that when fully recovered, the fish would be returned to the pet store to be sold to a new, happy home at full price. The Betta Hospital was born.

Every fish that entered the hospital seemed momentarily surprised. When I gently released the fish into their vases, they'd react with a sort of silent shock at being suddenly dumped into a very large amount of water, perhaps the most they had ever experienced. They'd see the plant and sometimes get excited, wedging their injured bodies between the soft branches. Sometimes they'd settle on the bottom and promptly take a nap. Usually within a day or two, they would start eating the freshly dried Bug Bites I offered—nutritious candy to a fish.

I wanted to avoid getting attached to the bettas. The goal was to bring them back to health and return them to the store. They were in my bedroom, but they were not my pets. At first, I refused to give them names. But at the insistence of my parents, I reluctantly began naming the fish in order of the Greek alphabet, in honor of the *B* letter for *beta*. Alpha, Gamma, and Delta were my first patients.

My parents became enthusiasts. The Betta Hospital was my special project, but my folks' growing love for it became a lifeline for me in the loneliness of our family trials. The hospital connected us, bringing us back together on the hard days. It literally brought my parents into my bedroom. From the hallway, I'd see them wander into my room, standing intently before my dresser and watching for the tiniest signs of progress: growth of an injured tail, a diseased fish's enthusiastic eating, the general happy waggling of fins. In the late evenings, there'd be a knock at my door. They wanted to tell me good night, but they also wanted to see the fish one last time, just in case one of them didn't make it until morning. Sometimes I'd come home from work, and a parent and family friend would be in my bedroom, gazing at the absurd-looking collection of tanks. My folks would be passionately explaining everything I'd taught them. "This is Epsilon. He has a bad case of popeye and clamped fin. And this is Delta—he also has clamped fin, but he probably has an extra bad bacterial infection from those wounds you see as well. And this is Mu . . ."

My folks and family friends developed personal favorites—fish that responded to their specific presence or that made unbelievable recoveries. After a fish had been fully cured, it was not uncommon for a friend to request to buy a specific fish they had come to know and love. The hospital fish were purchased not for their beauty or merely because someone had wanted a pet, but because the person had taken the time to really see that fish. To watch them. To know their personality. To admire their resilience. People were inspired by these fish. People loved their stories, so much so that I began to write a personal minibiography for each fish that returned to the pet store. The attached card would read something like this:

> Hello! I arrived at the store sick but have since made a full recovery in the Betta Hospital. I prefer big food pellets and enjoy blowing bubbles in my free time. I like to wave my fins when people visit my tank. For more information, please ask a store aquarist.

I was amazed by how quickly the hospital fish were embraced. They were always the first to sell. And what's more, the people who bought them would come to ask me questions. They were excited to know their fish. They wanted it to have a good, happy life. They wanted to know the best way to care for it. They thanked me for helping. Often they'd come back to the store weeks later to give me an update on their fish. These were my happiest days on the job.

Perhaps the saddest fish I ever saw was Pi, a pale blue-gray male who arrived with all his fins shredded from his body. He was a mere stump. There was a second fish in the triangular packaging with him, dead and decaying. Somehow, the two had been packaged together and on their voyage fought to the death. Pi barely survived. I brought him home purely out of mercy. I wanted to provide some comfort for his last day of life. I prepped his tank with antibacterial medicine. When I gently dropped him in, finless and hopeless, he sank like a thick leaf to the bottom and landed with a little thud. With no way to turn himself, he was going to be stuck sideways at the bottom until he breathed his last. I gently fingered him to lean against the base of the vase's wall, where he could breathe better for the moment. His gills opened and closed slowly. He wouldn't eat the tasty food offered. I told him goodbye that night, presuming it would be his last.

In the morning, not only was he miraculously alive, but he had also traveled across the vase's bottom to the far side and propped himself similar to how I had done the day before. I was awestruck. He ate two pellets and pooped. I did a daily water change and dosed his meds, watching his amazing technique for scooting himself across the bottom of the tank by the miniscule wiggle of his back end, much like a wounded caterpillar. My parents were mesmerized. We all celebrated his dramatic overnight improvement.

Over the course of a month, his nonexistent fins and tail grew to scraggily branches, and he began to swim again. He developed a habit of nestling into the plant, wedging his body between two branches to gently kiss the soft leaves with his blue lips. His bacterial infection cleared. His stubby fins continued to grow until they were long and flowing. His overall color brightened to a brilliant sapphire blue. He began to eat less in the evenings, which worried me until I caught my parents giving him "a snack" outside of his scheduled feeding time. He was their special favorite. And he had a strange affinity for my dad. Anytime Dad walked into the room, Pi would get active. If Dad put his face toward the tank, Pi would swim up to him and wiggle hard, as if bursting with joy. After making a full recovery, the time came for him to return to the pet store. I told my parents, and they immediately bought him. They also bought a five-gallon tank, stuffed it full of his favorite plants, and put him in the middle of the living room. Pi lived a long and happy life, dying of old age many years later.

At some point, we're all fish in the Betta Hospital. There are times when God intentionally sets us apart to heal, perhaps in the intimate isolation of a spiritual fishbowl. My season of preparing job applications late at night, processing through family trials and pain, and working at a pet store was a difficult time. It passed slowly. Most days, I would rather have been anywhere else. I struggled against the pain, crying out to God with questions. "What are you doing?" "Why am I here?" Looking back, I can see he was wielding a loving pocketknife. He was amputating the necrosis of my soul. He was treating the spiritual and relational infections within my family. He was providing the necessary environment and daily care required for my ultimate healing, for my restoration. He was mending my shredded heart and giving me a better story. He was growing the missing parts back and tenderly piecing together my fractures with gold, making kintsugi.

Even now, when I press into God, I know he is continuing to bring about complete restoration within me. We can be confident that the one who begins a good work within us will bring it to completion by the day Jesus returns.[5] The immeasurable joy of that day is thoughtfully touched on by an old Puritan prayer: "I live here as a fish in a vessel of water, only enough to keep me alive, but in heaven I shall swim in the ocean."[6]

Late one morning at the pet store, I was arms-deep in a dirty tank with a siphon hose in hand when my phone rang. I let it go to voicemail and didn't check it until the end of my lunch break, after

I'd finished my broccoli beef with noodles from a local Chinese restaurant. When I heard who it was, I called back in an excited frenzy. They asked for an interview.

Two weeks later I got a second call, which I rushed outside the store to take. On a patch of grass overlooking a parking lot by the noisy highway, I got the job offer I had been dreaming of, working for, and waiting for my whole life. It came not a moment too soon. When I got off the phone, I screamed for joy. A lady in a red coat taking her poodle in for grooming looked startled. My fingers dialed another set of numbers.

"Mom! Dad! I got the job!"

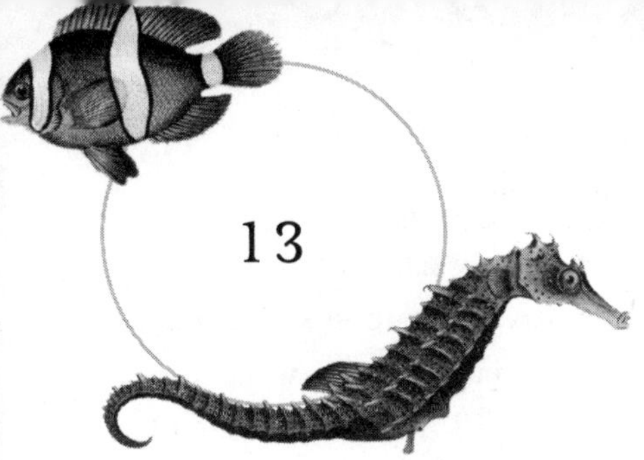

13

snuggling like seahorses

Early morning sunlight streaked down from Australia's Mount Stuart to touch the awning-covered units, teeming with tropical fish. As I walked by row after row of massive tanks, thick with cultured algae and bubbling from oxygenators, I smiled. The Marine Aquaculture Research Facility units (which were fondly dubbed "MARFU") were my happy place on campus. Between the daily grind of my graduate class schedule and hair-pulling exam preparations, I volunteered daily at MARFU, feeding the fish and scrubbing out red hair algae. My gravel-crunching footsteps could be heard up and down the buildings, as fish-smelling water dripped down my arms and my fingers gradually became pruney. After an hour or two of feeding and cleaning, I'd quicken my pace across campus to class, sneakers squeaking with moisture, salt water and sweat drying to my skin.

MARFU functioned as a wet lab for graduate students conducting research on the raising and keeping of saltwater fish. Filled with an assortment of units, both indoors and outdoors, the tanks housed multitudinous species, from miniscule Australian blennies to five-foot-long barramundi. Experiments with feeding schedules, nutrient enrichment, and reproductive behavior were constantly in progress. Although I didn't personally lead any experiments and wasn't even specializing in aquaculture, I helped multiple students with projects. The units were known for being messy, wet, and smelly, but I figured that any opportunity to be with fish was a good one.

I loved getting to know the characteristics and habits of different species, and especially the specific personalities of individual fish. There were hand-sized tangs and surgeonfish that were bold and friendly, taking food from my fingers without even bothering to raise their defensive dorsal fin barbs. There were the skittish blennies and gobies, nervously scooting their way around rocks to peek at me between strands of algae. There was one very gregarious lionfish, who squirted me with water every time I walked by his tank. He made a point of recognizing the people who were supposed to feed him, penalizing them for supposed tardiness. He would splash himself atop his tank's filter, pushing with his front fins to prop the top half of his body above the water. Then using his mouth, he'd shoot little jets of water. His aim was perfect, and I have yet to encounter another fish that surpassed this lionfish's innovative mischievousness and admirable gumption.

One of the more memorable projects I assisted with was a reproductive study on orchid dottybacks, a slim, finger-length, bright-magenta-colored fish. Since they were in high demand for the aquarium trade, aquarists were trying to figure out how to breed them. At the time, few people had successfully mated, hatched, and raised the species in captivity, which meant that most captive orchid dottybacks had been originally collected from the wild. This supply and demand problem had enormous implications for the long-term conservation of the species. We hoped that captive breeding would become the dominant source in the aquarium industry, leading to fewer orchid dottybacks being caught from the wild. But first we had to successfully breed the fish.

Our dotties were placed as male and female pairs in separate tanks and monitored. Nine pairs mated and laid clutches of tiny, translucent orange eggs inside a bit of provided PVC tubing, closed on one end to mimic their natural cave-like nest environments. It was my role to monitor the eggs for their viability, development, and continued existence. The male dotties stood guard over the eggs, curling their slender magenta bodies around the gelatinous orange babies. They were so intent on guarding their eggs, in fact, that they refused to eat. The protective fathers often remained inside the bit of PVC when I lifted it from the tank to check for eggs, but sometimes they'd become overly startled by the movement and jump from the tubing into the tank, waiting for me to replace the PVC in its original position, after which they would rush back to their eggs. Eventually, the eggs would fade from translucent orange to slightly colorless and opaque, which meant they were developing and closer to hatching.

But the greatest hindrance arose from what should have been the greatest asset. The eggs would sometimes disappear, gobbled up by a hungry albeit confused father who was overly intent on guarding them and subsequently starving himself. Unfortunately, most of the eggs were eaten, and I never saw any baby dottybacks hatch. The mystery of how to breed them remained unsolved during my time working at MARFU. But the good news is that, since then, several other organizations have achieved captive breeding for this marvelous species, providing greater protection for wild orchid dottybacks.[1]

I assisted on another project that was much simpler than the orchid dottyback one. I simply had to submerge a waterproof camera in a tank of lawnmower blennies, remove it after an hour, and download the footage for later review. The camera did all the hard work of capturing lawnmower blenny behavior. These relatively peaceful fish are hard to see, being mottled gray and green, excellent for camouflaging in their algae settings. They tend to skirt along the bottom of tanks, leaving a tiny trail in the sand as they go. The camera was intended to capture their reactions to specific changes in their environment, such as moving a rock or hideaway to the opposite side of their tank. We simply wanted to know what environmental arrangements the lawnmower blennies preferred. If they favored a particular arrangement, it would become our standard for keeping that species.

While I enjoyed monitoring clusters of orange eggs with protective fathers and recording a camouflaged fish's behavior, my favorite activity was feeding time. Like Pavlov's dogs, the fish knew when to expect food. On the hour, the water in their tanks appeared to boil. All their little lips lapped at the surface, fins creating ripples and slashes, anxiously awaiting the upcoming meal.

Feeding the baby clownfish was an experience in itself. Too young to identify feeding times, the pinkie-fingernail-sized orange-and-white babies would continue their slow migrations from one side of their vast one-hundred-gallon tank to the other. Each tank was filled with hundreds of baby clowns, looking more like giant globs of pollen than fish until more closely inspected. With the first sprinkling of food across the surface, the babies did not react, continuing on their steady course meandering around the tank. As the food began to sink, the top layer of fish began to gobble the tiny pellets. As food continued to fall, it was as if the heat was turned up. Hundreds of baby clowns rushed to the surface, lightly bubbling the water, frothing the surface with excitement. They would eat forever if I let them. And so, after several minutes of steadily adding food to the orange and bubbling tank, feeding time ended.

As much as I enjoyed the baby clownfish and all the other species I cared for, I could not help but have a favorite fish tank. In fact, I had a favorite fish. He had taken me by surprise, for amidst all the other marvelous creatures, I had not anticipated that I would fall in love with one in particular. He was the only MARFU fish I ever named. And he earned his name from the peculiar way he first approached me.

Between the lawnmower blennies and the water-squirting lionfish was another tank. It wasn't anything special, just another fifty-gallon tub containing water, algae, and fish. But the fish it held were delicate seahorses. Colored various shades of yellow, green, and gray, the seahorses grazed on algae throughout the day and, like all MARFU fish, were supplemented with a frozen blend of brine shrimp and bits of prawns, worms, and squid. They loved this meaty treat, slowly ambling out from the shadows to suck up flecks of food with their long snouts. Their tails curled with pleasure, and after eating, they promptly returned to their roosts among the seagrasses. All except one.

This seahorse came from the shadows with the others but swam straight to my hand. He did not eat immediately, opting instead to gently bump into my fingers, as if attempting a hug. The first time this happened, I was puzzled. Dropping the remaining food, I rotated my empty hand, palm side up. The friendly seahorse promptly swam onto my palm and curled up like a cat, laying his head on my fingers. I named him Mister Snuggles.

Mister Snuggles always swam to my hand. He would press his tail into my palm and gaze into my face. He would soak up the warmth of my hand as I enjoyed his affection. We would have a conversation about life in the tank and life in the classroom. I would feed him special little bits of food and stroke him with my thumb. He would stay there as long as he possibly could, until I pulled my hand from the water. On more than one occasion, it was because of Mister Snuggles that I arrived late to class. That fish will forever have my heart.

While there are more than thirty thousand species of fish alive in the world today (more than the combined total of all other vertebrates, including amphibians, reptiles, birds, and mammals), there is only one species of humans.[2] Species categorizations obviously do not account for the innumerable differences, quirks, characteristics, and attitudes that make up individuals. There is something to say of nature versus nurture, but whether we appeal to one or the other, our individual variances remain evident. Each of us was made special. Perhaps our spiritual stories are as diverse as our personal characters, with our responses to God potentially as different as fish in the sea, or in this instance, the fish at MARFU.

The world we know is a tank full of water. In it is everything we think we need. We were made for the ocean, but we find ourselves here, confined by four plastic walls and an uncrossable boundary, the water's surface. If we are fish, then God is the person taking care of us. His hand reaches into the tank to feed us and to look after the eggs we try to protect.

So often, we wrongly label God as a threat, not understanding that he plans to care for us rather than harm us.[3] If there is a threat, it is us; if eggs go missing, it is because we ate them. The responses available to an overly protective orchid dottyback are fight or flight. Fighting God works even worse than a dottyback fighting a person, and fleeing makes little difference, for God is omniscient and omnipotent. He sees the entire tank from his infinitely superior vantage and simultaneously stands within and beyond it. He knows us inside and out, being "familiar with all [our] ways" (Psalm 139:3, NIV). And so as an orchid dottyback may dread the appearance of its human caretaker, so some people dread the presence of God, ready to either fight or flee. For them, spiritual conversations can stir deep emotions, as they either aggressively oppose or eagerly avoid God.

On the other hand, lawnmower blennies wouldn't dream of fighting or fleeing. From the start, they do everything possible to elude the one who cares for them. They blend, hide, and hold very still, hoping to go unnoticed, terrified of what it would mean to be spotted. People whose spiritual lives resemble the behavior of a lawnmower blenny may outwardly appear apathetic to God and spiritual things. But perhaps the very act of camouflaging—of trying to hide their existence from the one who brought them into existence—indicates a very real care indeed. They take no risks and ask no questions, thinking it far safer to hide and be wrong about God than to emerge.

Baby clownfish come churning through the water to be cared for by the hand that feeds them. They are excited and overwhelming. But they care only for the hand, knowing nothing and caring nothing of the person to whom it is attached. They are oblivious, thinking only of the good things that fall from the hand, how it can fill their stomachs and satiate their ravenous desires. They search after every good thing that falls from the hand but miss the whole person. God is only a vending machine to those who live like baby clownfish. They must be given their three square meals a day, or else they will turn and follow him no more.

Finally, there is Mister Snuggles. Truth be told, he is not a normal fish. He does not view the hand that feeds him as a threat. He is not ruled by fear or apathy. He sees the hand for so much more than merely what he can receive from it. He goes to the hand, knowing that it is gentle and warm and attached to a person. He looks at the person. He snuggles down and stays. He knows he is loved and safe.

In some seasons of life, I have been like an orchid dottyback, misunderstanding the hand that cares for me as a threat. At other times, I have been the lawnmower blenny, apathetically hiding in the shadows, refusing to engage. I have also been a baby clownfish, faithlessly making a vending machine of my Provider and missing out on the greatest good that could ever be given to me. I wish I was as unwaveringly loving as a certain seahorse.

Mister Snuggles responded to me the way I long to respond to God, the way that we all ought to respond to him. When we understand and remember who God is and what he has done, specifically for us, we are compelled to draw near to him. We want to know if he is who he says he is. We want to make sense of him. We long for just a moment of his presence. We begin to crave relationship with him. When we realize the cosmic story, how deeply God loves us, the sacrifice of Jesus, and the intimate relationship with him now made possible, we will want to get as close to the heart of God as we possibly can. And because of what he has done, we can get that close.

In life, it often feels as though we are swimming in a tank of four plastic walls and an uncrossable surface. "We are strangers here,"

observes C. S. Lewis. "We come from somewhere else. Nature is not the only thing that exists. There is 'another world,' and that is where we come from. And that explains why we do not feel at home here. A fish feels at home in the water. If we 'belonged here' we should feel at home here."[4]

Perhaps we are not truly fish after all. Perhaps the world is not merely this bin full of water. Our lives can feel dissonant and confining, and yet it is still possible for us to experience the hand of God, penetrating through every boundary to hold us. We need only go to his hand. As we gaze into his eyes, we see in him the promise and hope of someday seeing the ocean with our own. We hardly comprehend what the ocean is, for we have never experienced it, though we were made to swim there. And we were made to swim alongside him.

God makes this possible. He does more than merely watch us. We can rest in his presence because he is present. We can hold his hand because he extended it to us. We can respond to him because he pursued us. We can love him because he first loved us.[5] God invites us into relationship with him. It is his pursuit and our response that forms this relationship, that shapes our future together or apart. He has reached exceedingly farther toward us than any person feeding fish at MARFU.

He gives each of us the opportunity to know him personally. We can know him through both his Word and world—the words he has given us in the Bible and the world he created around us. We will surely find him when we seek him with our whole hearts.[6] Longing to be seen, known, and loved, may we each recognize the seeing eyes, the knowing mind, and the loving heart of the one who reaches out to us. May our outward responses be accurate reflections of what we internally seek, that we will get as close to the heart of God as we possibly can, snuggling like seahorses.

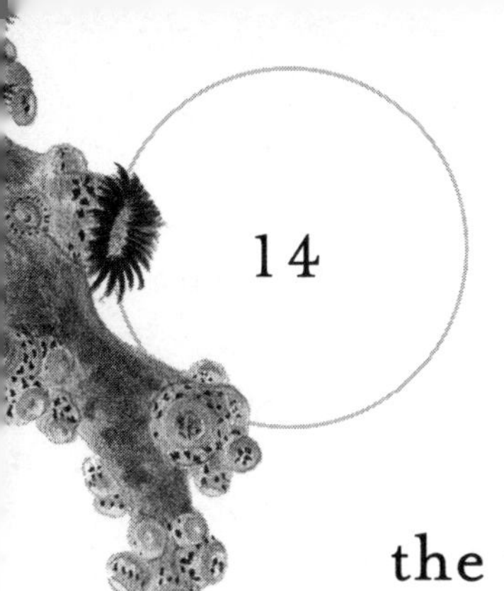

14

the promised garden city

The bird was drowning. I knew on sight that this shorebird was meant to hunt and stalk along the waterline, not to be caught up in twenty-something feet of water and swept from shore. One of its slender wings was bent backward at an abnormal angle, lifeless weight sinking in the water. Its other wing flapped, splashing violently in a futile effort to free itself. The poor creature panted with exhaustion, blinking back salt water from its eyes with an expression of tired terror. Something needed to be done.

A long-handled fishing net and five-gallon bucket did the trick. Because the bird was just out of reach of the handle's length, it was a process to coax it into the bright orange bucket. Wild creatures are wired to resist the unnatural and superficial, sometimes at the cost of their lives. Despite

all my gentle encouragement, it was not persuasion but exhaustion that drifted the drenched bird within scooping range of the bucket.

Now that the helpless creature was in my firm, gloved hands, I could see several things. It was a female brown noddy, likely a year or two old, come to the island to build the first nest of her very own. But all the hope of motherhood had left what would otherwise have been her brave, bold expression. With only one hesitant nip at my fingers, she blinked a look of helpless release, a fighter forcibly surrendered to conditions rendering her powerless. And yet she no longer looked afraid as she had in the water. Whether from pure exhaustion, mortal resignation, or some greater peace, the young brown noddy closed her eyes as I gently stroked her head.

There is little hope for a bird with a broken wing. And yet broken birds always remind me of the Emily Dickinson poem:

Hope is the thing with feathers
That perches in the soul,
And sings the tune without the words,
And never stops at all.[1]

It is the same with birds, even as they stop breathing. But the brown noddy I held did not stop breathing, and I knew just where to take her.

Years ago, the Audubon Society funded a "fountain" on Garden Key in Dry Tortugas National Park. It was little more than a metal tray, surrounded by a bed of brick, with a hidden hose attached to a spurting water pump that needed to be unclogged all too often. A glorified birdbath, if you will. But it's true that the birds did glory in it!

In the Dry Tortugas, most shorebirds are visitors. They come for a season and leave the next, winging their way across vast distances. In winter, pelicans with bulbous beaks and thick, oar-like wings are the norm, flying low in search of fish and happily face-planting into the ocean when they find some. Come springtime, you can spot the occasional burrowing owl, shyly popping up from its underground nest to catch a mouse and feast in the branches of a nearby sea grape tree. Masked booby birds can be heard clownishly squawking and strutting the beaches of Hospital Key year-round. In autumn, sections of the keys adjoining Fort Jefferson are closed to safeguard the largest breeding colony of sooty terns and brown noddies found in the continental United States. These birds thrive in tropical island ecosystems, each filling an important niche.

The fountain was the birds' best source of fresh water on the island, located in a little clearing of soft grasses, surrounded on all sides by a tall fringe of bending buttonwood trees. It was a peaceful spot and, I figured, the best place on the island to die. If I were a bird with a broken wing, I would want to be laid to rest at the fountain.

As I set the limp brown noddy down and released her from my hands, she opened her eyes. She did not look at the gurgling fountain or the soft grass or the bending buttonwoods rustling in the breeze. She looked at me with a knowing realization that I would help and not harm her, that she was, in fact, safe. And then she closed her thin eyelids over her bright eyes and fell fast asleep.

The bird that slept by the fountain disappeared early the next morning. But later that day, a young female noddy was observed with one wing drooping to the ground, happily frolicking among flocks of other brown noddies by the shoreline, as they scratched their nests into sand.

The Genesis creation account is usually read for the purpose of understanding how creation unfolded under God's direction. He made all things "good"—a specification differentiating the biblical creation story from other ancient Near Eastern creation accounts. But there is something else. The Genesis story reveals the purpose of humankind and all of creation: We are meant to glorify God and reveal his character.

God gave humankind a specific mission in the Garden of Eden. "The Lord God took the man and put him in the Garden of Eden to work it and take care of it" (Genesis 2:15, NIV). This verse serves as a mission statement for humankind on earth, informing us of our originally designed role in creation designated by our Maker. But unfortunately, a bit of nuance was lost in the translation of Genesis from Hebrew into English.

The original Hebrew for "to work it and take care of it" has also been translated to "to till it and keep it" (RSV). But perhaps the most meaningful translation of this verse is "to serve and protect."[2] Humankind was placed in the Garden to serve and protect the rest of creation. How different this sounds from the typical interpretation of Genesis 1:26: "Let us make mankind in our image, in our likeness, so that they may rule over the fish in the sea and the birds in the sky, over the livestock and all the wild animals, and over all the creatures that move along the ground" (NIV). In one place humans are called to serve and protect; in another, to rule. We might pause to ask, why this discrepancy?

We must remember that God's definition of "ruling" is not that of this world. God does not rule by terror or an ironclad fist. He is not a democratic president, a totalitarian dictator, or a parent spoiling selfish children. Rather, he rules perfectly. He reigns from a place of perfect justice and perfect love. And in his original design, humankind was called to rule likewise. This is how we can reconcile the "ruling" of one verse with the call to "serve and protect" creation in the other. Pope Francis puts it like this:

> We must forcefully reject the notion that our being created in God's image and given dominion over the earth justifies absolute domination over other creatures. The biblical texts are to be read in their context, with an appropriate hermeneutic, recognizing that they tell us to "till and keep" the garden of the world (cf. Genesis 2:15). "Tilling" refers to cultivating, ploughing or working, while "keeping" means caring, protecting, overseeing and preserving."[3]

What a difference there is between our original purpose on earth and that which we so often fulfill! Since the introduction of sin into the world, humans have continued to perpetuate a cycle of suffering and death, all the while yearning for a balm to the pain we were never meant to experience. Even in this longing place, the sinful world where we can no longer reach the eternal fulfillment for which we were designed, we are still meant to live out our original purpose.

Through the saving power of Jesus, through the phenomenal cosmic story that God has written on and over and through all of creation, our purpose now transcends the kingdom of earth to reach the Kingdom of God, a reality that overlays all of earth and everything beyond, even out of our sight. "From the moment of humanity's first sinful choice, God set in motion a promise to redeem, restore, and make all things new. Sin's power has an expiration date."[4] While in this sinful world, this land of the living, "we are called to be instruments of God our Father, so that our planet might be what he desired when he created it and correspond with his plan for peace, beauty and fullness."[5] As we become attuned to the heart of our Creator and learn of his intention for creation (including us), we can take hope in the fact that the original design of earth now teaches us of heaven, the Kingdom of God into which we will someday be completely restored.

The kingdom of earth might best be described by a concept from theoretical ecology known as the tragedy of the commons.[6] When a space or resource is commonly available to people, that space or resource can become tragically overused.[7] Instead of people considering how to serve and protect the space or resource they have in common, they try to beat other people to it. In a vicious cycle, people compete until what is available is either severely diminished, damaged, or dead; everyone loses in the end. Voluntary restraint, in other words, is not merely an individual decision but rather a collective choice either by agreement or (more likely) disagreement when one person's restraint is supplanted by another's selfishness. One example of this tragedy is our oceans, where international waters are zones of extreme overfishing and pollution dumping one-upmanship.

In our kingdom of earth, people share a common home not only with one another but also with the rest of creation. The tragedy of the commons is one ecological example of our failure to carry out God's original purpose for us. When it comes to creation, we are far better at using and enslaving than we are at serving and protecting.

This could not be more against God's original design for humankind, which was created in his image, his likeness, to rule in a manner that reflects his just and loving character.[8] But what does "in his image" truly mean? There has been much scholarly debate on the specifics of this phrase's meaning, but some things are certain. Humans were made intentionally by God to be a reflection of himself to the rest of creation. Simply put, "we are His image on earth."[9] While this image is misrepresented in our decision to sin, we are reminded of God's promise for those who choose to know and love Jesus. "If anyone is in Christ, he is a new creation. The old has passed away; behold, the new has come" (2 Corinthians 5:17, ESV). That is to say, by choosing Jesus, we are no longer the misrepresentation of God's image that sin has led us to become, but rather becoming the recreated version of his original design. We are being remade in his image.

Just as humans have failed to fulfill our originally intended purpose, creation itself falls short of God's original design. We know from the Genesis story that this is entirely because of human choice. Instead of serving and protecting, we led all things astray. We introduced sin into the world, resulting in the pain, suffering, destruction, and death experienced by all creation. The ecology of the Garden of Eden was originally beautiful. Now only old gleams of that beauty remain. Without accounting for this, we will never begin to grasp the fullness of God's promise for the collective future of humankind and the rest of creation.

Since all creation was made to glorify God, this is exactly what we will do when his Kingdom has finally been restored. The best part is that we will again experience the intimacy of relationship with him. Humans were always designed to be with God, just as we were designed to reflect him. In his restored Kingdom, we will finally do both!

Although death will ultimately be defeated, it continues to ravage our world and destroy our earthly bodies. The apostle Paul explains the reason for this using a metaphor from horticulture: "What you sow [the body] does not come to life unless it dies. When you sow, you do not plant the body that will be, but just a seed" (1 Corinthians 15:36-37, NIV). This life must end for the next to begin. Spiritually, this has already occurred for those who have chosen salvation from Jesus. We were spiritually dead in our sins; now we are alive in Christ. But physically, this transformation has yet to occur. Jesus conquered death, but death itself has not yet been eliminated.

So what will it look like to be physically restored? As a lifeless seed must be placed in the earth, similarly we must return to the earth before we can be transformed into the physical creation that will be our body for our eternal life. Restoration is a biblical promise, one made directly to us by God. "The God of all grace, who called you to his eternal glory in Christ, after you have suffered a little while, will himself restore you and make you strong, firm and steadfast" (1 Peter 5:10, NIV). We can know from this promise that just as to be in Jesus is to become a new creation,[10] to be restored by God is to be made strong, firm, and steadfast. We can also know that restoration involves being called into his eternal glory. While nobody yet walking in the land of the living has been able to provide a definition for what this might be like, there are numerous accounts in the Bible (Isaiah, Ezekiel, Daniel, and Revelation, to name a few) of people doing their best to describe the glory of God, giving us a miniscule, fractional understanding of what the reality will be. But as for our physical bodies, we are promised that they will be resurrected.

While the nature of resurrection has been debated, we know that Jesus' resurrection serves as a template. It was both a spiritual and a physical event. As Adam led the world into sin and Jesus has provided a way of salvation out of that sin, a comparison between Adam and Jesus extends to our resurrected bodies as well. "Adam, the first man, was made from the dust of the earth, while Christ, the second man, came from heaven. . . . Just as we are now like the earthly man, we will someday be like the heavenly man" (1 Corinthians 15:47, 49).

> Jesus' triumph on the cross finished the work of our salvation. His death atoned for it. His rising proved that sin and death have no power over the life-giving God, no power over Jesus, and no power over us. Though we will still die on this earth, our lives in Christ begin now. . . . Because of Christ's victory, we go from life to life.[11]

Both spirit and body will be resurrected. Jesus resurrected into a physical body that could be touched, that showed marks from his earthly pre-resurrected life, and that felt hunger for food.[12] His physical body was also capable of very unearthly things, being able to disappear and reappear much like a ghost.[13] He no longer appeared constrained by the biological and physical properties that rule the world. As such, we can suppose that our resurrected bodies will be similar in form and function. As Timothy Keller said, "God created body and soul. He's redeeming body and soul. That's what the resurrection means . . . and He's going to restore it."[14]

Interestingly, creation is used to describe the difference between our earthly bodies and our resurrected, heavenly bodies.

> Then God gives it [the seed] the new body he wants it to have. A different plant grows from each kind of seed. Similarly there are different kinds of flesh—one kind for humans, another for animals, another for birds, and another for fish.
>
> There are also bodies in the heavens and bodies on the earth. The glory of the heavenly bodies is different from the glory of the earthly bodies. The sun has one kind of glory, while the moon and stars each have another kind. And even the stars differ from each other in their glory.
>
> It is the same way with the resurrection of the dead.
>
> 1 CORINTHIANS 15:38-42

Our heavenly, resurrected bodies will differ in unique, wonderful ways from our physical, earthly bodies. However, will humans be the only part of creation to resurrect? Could it be that this resurrection extends to the rest of creation as well?

I believe the only basis for the argument that humans alone resurrect is an overreliance on a presupposed meaning of humans having been made "in God's image." There is a pattern throughout Scripture from the Garden of Eden in Genesis to the Garden City in Revelation, from God's provision for the animals on Noah's ark to Jesus' comments about falling sparrows. God's love for creation is made clear in the Psalms and in this wisdom offered in Ecclesiastes: "Who can prove that the human spirit goes up and the spirit of animals goes down into the earth?" (3:21). Some people claim that animals do not have spirits; I would like to know how they would answer the question posed by the author of Ecclesiastes. Whatever the case, we should consider the ecology of the resurrected world and eternal life that will be our inheritance. We might pause to ponder all the potential inhabitants of the Kingdom of Heaven.

The concept of restoration applies not only to the physical and functional roles of humankind in creation but also to how we relate to God. He made us for himself, as a part of creation specially designed to reflect him but particularly so that he could be in relationship with us. This is evident from the wording of the Genesis creation account, where descriptions of God's intentionality in creating humankind showcase his love and triune nature. He also took care in creating both man and woman for each other, thus emphasizing the importance of relationship and togetherness. We read that Adam and Eve heard God walking in the Garden. The casual details offered make clear that this was a familiar event. The good news of what Jesus has done for us best radiates God's desire to be in relationship with humankind and the rest of creation. "For this is how God loved the world: He gave his one and only Son, so that everyone who believes in him will not perish but have eternal life. God sent his Son into the world not to judge the world, but to save the world through him" (John 3:16-17). It stands then that in the restored Kingdom of God, the world will return to restored relationship with God.

Like humans, the rest of creation receives its own promise of restoration. "For God was pleased . . . through him [Jesus] to reconcile to himself all things, whether things on earth or things in heaven, by making peace through his blood, shed on the cross" (Colossians 1:19-20, NIV). All things are reconciled to God through Jesus. "In a world groaning beneath the weight of sin (Romans 8:22)," says Phylicia Masonheimer, "we rejoice in *hope* that Christ came not just to save us eternally and restore us to God but to restore *all things*."[15] Just as God so loved the world that he sent Jesus, he also so loved the world that he has fulfilled a plan to redeem and restore everything. "And the one sitting on the throne said, 'Look, I am making everything new!'" (Revelation 21:5).

Skeptics may pause to wonder, why go to the bother of recreating? If God went to all the effort of the Genesis account's description to make everything in creation, if he designed and implemented physical and evolutionary processes to make everything in existence, why do it all a second time around? Why reinvent the wheel? Why not recycle a bit?

The basis of these questions is a common misconception. "This is because, contrary to the popular saying, heaven is *not* our home. Earth is. Not Earth as it is now, but Earth as it will be in the future. Our hope isn't for another *place*, but another *time*. . . . Jesus called it 'the renewal of all things.' Paul called it 'the kingdom of God' and 'eternal life.' Peter called it 'the time . . . for God to restore everything.'"[16] The wording of God's declaration in Revelation 21:5 matters. He says, "I am making all things new" (ESV), not "I am making all new things."[17]

The apocalyptic destruction that precedes this restoration is not merely the end of all things, but rather the end of all things in the form and function that they used to know, the version that had been impacted, warped, and misshapen by sin. It is significant that God's declaration of "I am making all things new" is followed by a phenomenal description of a brand-new and wonderfully strange ecology, such as has never been seen before.

The Kingdom of Heaven is described by John Mark Comer as "a city with walls and gates and streets and dwellings and a river and a forest and *culture* . . . draped in language straight out of Genesis."[18] There is a spring of the water of life, also described as a "river of the water of life, as clear as crystal, flowing from the throne of God and of the Lamb down the middle of the great street of the city" (Revelation 22:1-2, NIV). There is a kind of tree described in the same passage as "the tree of life," which has the same name as the tree in the Garden of Eden that granted eternal life,[19] growing on both sides of the river. I had always assumed while reading this description that there were probably multiple trees of the same tree of life species growing on either side of the river, but someone once posited a true botany question to me: What if there is somehow one singular tree growing on both sides of the river? We will have to wait and wonder and see. What we do know is that God will continue to fill everything with his power and love:

> The ultimate destiny of the universe is in the fullness of God, which has already been attained by the risen Christ, the measure of the maturity of all things. . . . All creatures are moving forward with us and through us towards a common point of arrival, which is God, in that transcendent fullness where the risen Christ embraces and illumines all things. Human beings, endowed with intelligence and love, and drawn by the fullness of Christ, are called to lead all creatures back to their Creator.[20]

I challenge you to read the description of the Kingdom of God from the perspective of an ecologist, from the perspective of someone who can see that God loves his creation. Dream about what the throne of God would be like as a waterfall. Wonder at whether the water of life will be teeming with living things. Imagine the tree of life as a clonal banyan tree, now as a series of deciduous fruit trees, now as a jungle-tangle of vines radiant with the glory of God. Picture the Kingdom of God as something beyond your initial imagining. Consider what it might mean to dwell in the promised Garden City.

As a marine biologist, I must address one minute detail. The apostle John's description of the Kingdom of God includes this: "Then I saw a new heaven and a new earth, for the old heaven and the old earth had disappeared. And the sea was also gone" (Revelation 21:1). To the ears of an ocean lover, this solitary verse initially screams utter betrayal. Why would the Creator of the sea and everything in it abolish it from his promised Garden City where all things will finally be restored to himself?

As an ecologist, I find it contradictory that God would care little for the sea and yet care deeply for the rest of his creation. As a marine biologist who was called by God to this particular career path, I find it contrary to believe that the job God called me to in earthly life will not also be restored in some capacity in his eternal Kingdom. If trees and rivers and "all things" matter to him and will be restored, will not also the corals and seahorses and brown noddies? Intensely motivated to find the true answer, I landed on the following explanation.

Throughout the ancient world, and particularly throughout

the Bible, the sea was used as imagery and as an allusion to chaos, disorder, fear, and destruction.[21] The pre-created Genesis world, despite being called void and empty, contained water.[22] Why would a place that is truly empty also contain water, unless that description held a layer of symbolism? Additionally, one of the most famous miracles of Jesus was when he calmed the storm over a body of water, demonstrating his authority over even wind and waves.[23] I believe this story is both an actual event and a spiritual metaphor, demonstrating that Jesus has authority over all things, including the literal climate and turbulent storms of our lives.

I do not know if there will be a literal sea in the Kingdom of God. I hope there is, I deeply suspect there is, and I anticipate that there will be. But there will not be any metaphorical sea in the Kingdom of God; the tumult and carnage that the sea represented in ancient times will be gone. It will literally be "no more" (Revelation 21:4). Death and all its ungodly companions will be utterly defeated. Of that sort, there will be no sea.

The air is thick with praises to God—songs with luxurious harmonies from "every tribe and language and people and nation" (Revelation 5:9). The singing flows from "every creature in heaven and on earth and under the earth and in the sea" speaking a secret but innately understandable tongue of glory, glory to God for ever and ever.[24] In a place beyond the limits of time and space, we come amidst joyously singing throngs to kneel at the foot of the watery throne, radiant with rainbow fractals of light pouring from the presence of the Lord God Almighty.[25] Humankind gathers with the rest of creation to sing and sip the sweet, flowing water. We garden and harvest and eat from the trees, savoring the presence of our Creator, so deeply intimate beyond the reach of any words to be written.

I dream of these moments, the coming of this place to the created world I love, finally restored. In the interim, I'll live like naturalist John Muir: "As long as I live, I'll hear waterfalls and birds and winds sing. I'll interpret the rocks, learn the language of flood, storm, and the avalanche. I'll acquaint myself with the glaciers and wild gardens, and get as near the heart of the world as I can."[26] Because here on earth, I can catch those little glimpses of the Kingdom of God, seeing aspects of his character written across the story of creation. Like the joy of seeing a brown noddy frolicking from a fountain to sing over her little nest, these gleams of beauty remind me of the promised Garden City, a place where both the broken hearts of people and the broken wings of birds will be completely restored.

15

bubble-brained

I was lost, slowly turning in circles while scanning the blue mist with wide eyes, scouring the silty distance for my dive buddy. I'd been diving for over a decade and had never been lost underwater until now. After the first few moments of realization, I glanced at my pressure gauge. There was still plenty of air left in my cylinder, which was both relieving and frustrating. Relieving, because there was no threat of being low on air. And frustrating, because it meant that if my buddy and I were not reunited soon, I would have to end the dive and resurface, hoping to find my buddy on the boat. Any diver worth their salt hates to end a dive early.

That said, it was time to implement the lost buddy procedure. I looked at the dive computer on my wrist, a fancy piece of technology that tracked important details of my dive: notably time, depth, and safety stops. I quickly set it to a one-minute

timer and continued turning in circles. I had become vaguely aware of the strong current, nudging me farther from the spot where I'd first realized I was lost. According to the compass on my wrist, I was drifting southeast. I turned my body into the current and began lightly kicking against it. To keep from losing my place, I trained my sights on a ginormous coral, attempting to hover buoyant at a fixed point above it. But it was difficult to do this while also turning in a circle, kicking against the current, and visually searching for my dive buddy. All the while, a sharp metallic clicking filled the water in the familiar rhythm of someone rapping on a door, *dut, dut-dut, dut, dut . . . dut, dut!* Although sea creatures make many sounds, none of them are metallic. Of course, the noise was coming from me. I tapped the rhythm behind my back on my aluminum scuba cylinder with a double-ended stainless steel clip grasped tightly in hand. Although sound is nondirectional underwater, I hoped that my buddy might hear the clicking, recognize I was missing, and retrace his fin kicks back to me.

When the timer on my wrist vibrated the end of its minute, my dive buddy was still nowhere within sight. I'm not going to lie; this definitely made my breathing rate go up a bit. I could feel my heart rate pulsing behind my ears. There is something unnerving about realizing you're totally alone while underwater in the open ocean. Nobody is coming for you. If you have a problem, it's on you. It's these sorts of scary moments that reveal your ability to make quality decisions in a state of fear. They unveil courage or the absence thereof. They also test the level of planning on the part of a dive team. And planning is critical for diving.

I've heard it said that a diver is only as good as the sum of their training and gear. Without a plan, I wouldn't have known what to do. I might have sat aimlessly spinning in circles above a big coral until I got dangerously low on air and even more frightened.

Or I might have immediately gone to the surface, not even trying to look for my dive buddy. I might have left him alone underwater, frightened and uncertain, while I made an immediate exit for myself. The plan is what kept us aligned, even as we were drifting apart. It didn't altogether prevent the problem of us being separated underwater because unexpected things (like a strong current) happen, but the plan meant that there was a step-by-step process to follow when things went sideways.

I knew what to do. The planned minute of searching was up, and now I unzipped the pocket on the right hip of my BCD. There were numerous goodies inside: a waterproof notebook and crayon, a bundle of multicolored zip ties, a soft mesh bag for carrying extra things, a spare dive compass, a stainless steel reel wrapped with over a hundred feet of neon orange cave line, and a safety sausage (also called a surface marker buoy or SMB). I was after the reel and safety sausage. I pulled them out and, quickly stowing the rest of my gear and rezipping the pocket, turned my attention to the two items.

The orange cave line was attached to the side of the metal reel. An extra double-ended clip, like the one I'd used to tap my cylinder earlier, kept it from unraveling. I unclipped it and attached the line to a metal loop on the safety sausage, which was a four-foot-long sausage-shaped pocket of bright yellow and orange, with reflective tape on each side. On one end, it had a valve for inflation and, on the other, a pressure release valve. Safety sausages function like underwater balloons. Because air is less dense than water (the reason bubbles travel up to the surface), a diver can inflate the balloon with their exhales or regulator, then let the sausage fly up to the surface.

I was going to do a sort of inverse-fishing method to locate my dive buddy. I would attach the orange cave line of the reel to the safety sausage and inflate it so that it floated to the surface while I (like a caught fish) held on to the reel below. Once the safety sausage reached the surface, I would slowly ascend while rolling up the line in my hands, essentially reeling myself in. This way, my location in the water was marked on the surface, lowering my risk of being accidentally run over by a boat. And the bright color of the sausage would help our boat's captain spot me from afar, as well as alert my team that something on the dive had not gone according to our original plan. If my dive buddy followed our "lost buddy procedure" too, he would do the exact same thing. The boat captain would see two safety sausages on the surface, and my dive buddy and I would both be collected by the boat.

The plan worked perfectly, and within a few minutes of inflating my safety sausage, my dive buddy and I were reunited on the boat. It wasn't until we debriefed and discussed the dive that I recognized the sheer number of decisions that I had made from the moment I realized I was lost to the moment we were reunited. It was evident how much more lost I could have become if I had chosen to deviate from the plan, or if I had not had the training and gear necessary to complete it. Because we had a solid mutual understanding of what to do beforehand, along with all the right training and gear, I knew what decisions to make as the moment for each of them arose. My breathing underwater might have increased in the fear of the moment, but because of our plan, training, and gear, I was never truly lost.

Decision-making in life operates in a similar way to that of the "lost buddy procedure." Things tend to go best when we follow some sort of framework for how decisions will be made. For those who know and love Jesus, this is particularly the case. The good news of who God is and what he has done for us and all of creation impacts every aspect of our lives. So much so, that it can and should be very evident to the people around us. Knowing and loving God functions for a Christian the way that good training functions for a scuba diver. It teaches us. It trains how we see and respond to the world. It is represented in the decisions we make—even the smallest ones.

But that doesn't mean that decision-making is easy any more than having a map of a dive site and a compass on your wrist mean that getting to the marked location will be easy. The swim might be hard. Unexpected currents may pull you off course. Thick, underwater hazes can roll in and make visibility challenging. Big, startling fish may distract you from the route ahead. Despite all the best strategizing, sometimes you get separated from your dive buddy and have to interrupt your original dive plan to implement the lost buddy procedure. Just so, life has its challenges, and we make mistakes. But in the same way that planning, training, and donning the right gear make all the difference to a diver, three things make all the difference in decision-making for those who desire to know and love God.

The first might sound familiar because we have mentioned it before. The Bible is "God-breathed and is useful for teaching, rebuking, correcting, and training in righteousness, so that the servant of God may be thoroughly equipped for every good work" (2 Timothy 3:16-17, NIV). Given our access to the very Word of God, the question we ought to ask ourselves when making decisions is whether or not we are exposing ourselves to Scripture. Quite simply, are we becoming familiar with God's words? If the Bible is God-breathed, are we metaphorically letting God breathe on us? Are we spending time listening to what he says?

Spending time reading the Bible is the Christian's equivalent of a diver knowing how to use their scuba equipment. It's like glancing down at a pressure gauge to see how much air is left in the cylinder on your back, making sure you have enough to breathe. It's like keeping a dive compass on your wrist so that if you drift off course, you have a tool to help you get back on track. It's like having a safety sausage in a pocket that you turn to in emergencies. When things don't go to plan, all those tools are there with the potential to direct the next step. Divers know each item, where it's located on their body, and how to best use it. They have practiced and trained with their gear so that they are equipped to do whatever sort of work may be required of them. A Christian should desire to do the same, proactively investing effort in understanding the Bible so that they are equipped "to do good works, which God prepared in advance" for them (Ephesians 2:10, NIV).

The second thing that helps with decision-making is wisdom. The Bible describes wisdom as being of great worth in this way:

> How much better to get wisdom than gold,
> and good judgment than silver!
>
> PROVERBS 16:16

Personally, I wish I had more wisdom and the discernment that accompanies it, particularly in my career as a marine biologist. Both have saved me from many a close scare, but they could have saved me from many more had I been in greater possession of their riches.

As a kid exploring the northwestern Pacific coast with my family, I would stuff my pockets full of shells, studying the husks of whelks and the hinges of scallops. I celebrated the victory of an intact sand dollar shell, the little bones rattling around inside its thin cookie shape, stamped with the imprint of a looping star. Once I spied the remnants of a plastic bag washing up in the thinning surf. White, sloppy, partly filled with sand, it was the perfect vessel to hold the shells that could no longer fit inside my soggy coat pockets. When I reached down to scoop up the bag, my fingers came into contact with its congealed surface, and as I began to lift the strange piece of litter, I recognized that it was not, in fact, a plastic bag.

In the shallow water, a thick translucent tentacle slipped off its cap toward my thumb. Abruptly, I dropped the jellyfish back on the beach. A wave washed over the dead creature, flipping it upside down and revealing a flowerlike spread of curlicue tentacles, which were being pounded to pieces in the slow surf. The tentacles were lined with stinging cells, called nematocysts, which remain active even after the animal dies. I'd had enough wisdom to drop the jellyfish before it stung me, but not enough to recognize that the bag I sought was not a bag to begin with.

One of the most wonderful things about wisdom is that we can ask God for more of it. "If you need wisdom, ask our generous God, and he will give it to you. He will not rebuke you for asking" (James 1:5). We aren't meant to simply garner wisdom on our own. True wisdom given by God can't be self-manufactured because it is "first of all pure. It is also peace loving, gentle at all times, and willing to yield to others. It is full of mercy and the fruit of good deeds. It shows no favoritism and is always sincere" (James 3:17). We need God to access those qualities, to obtain true wisdom.

Practically speaking, prayer is a way in which we can ask for and receive wisdom. As we consider decision-making from the place of wanting to know and love God, we ought to ask ourselves—just as we did regarding the Bible—Are we praying? Are we engaging in conversation with the one whose foolishness is wiser than human wisdom, and whose weakness is stronger than human strength?[1] As we do, we are acting like divers who have a plan and communicate with their dive buddy in advance of the underwater mission. As we make decisions, we should desire to be on the same page with God.

The third thing that helps with decision-making is having a dive buddy. Safety-conscious divers tend to operate in pairs for the simple reason that "two people are better off than one" (Ecclesiastes 4:9). But perhaps an even better example of how relationships can help in making decisions comes from canoeing. If you have ever sat in a canoe, you probably know how easily it can tip and flip over. Many a canoe has capsized early on when somebody first climbed aboard. Lean too far over an edge, tilt the boat too far, and the whole thing either begins to fill with water from the side, or it somersaults upside down and dumps its human contents into the water.

The distribution of weight in a canoe is important for balancing and efficiency while paddling. Similarly, paddle placement along the front and back of the canoe has a large impact on its overall movement. Canoeing can be tremendously fun, whether the boat tips or not, but canoes work best when there are two people in them, one at the front providing speed with their paddling and another at the back guiding direction with their strokes. Without the person in the front, the canoe progresses very slowly. The person in the back has to constantly switch hands to paddle on both sides of the boat, trying to maintain forward momentum with little luck because all of the weight is in the back. Without the person in the rear, the person paddling in front struggles to turn the canoe. The boat drifts, often in directionless circles. The person in front spends much of their energy getting nowhere in particular.

When making decisions, particularly important ones, we ought to think of the path ahead like the route of a canoe. Just as the weight distribution of people matters in the boat, having trusted people provide input to an important decision also matters. In some ways, it might seem so much easier to go through life living in isolation and "paddling our own canoe." But inviting the wise input of trusted people is the equivalent of welcoming the strength and position of another paddler to your boat, aiding in forward movement, overall efficiency, and direction while turning. All that to say, navigating hard decisions is best done with input from community. And this community, like a canoe, provides different positions from which we can love each other.

I've heard divers occasionally excuse or justify their behavior underwater as the result of being bubble-brained. What on land is a term for describing an incredibly unintelligent person is used underwater to name the bubble-like confusion that strikes when having to make clear-cut decisions while suspended in a watery environment so foreign to our own. The sharp taste of salt in the regulator, the blinking back of sweat dripping into one's eyes behind the mask, and the harsh hiss and subsequent gurgling of a stream of bubbles released with every breath—these nuanced details of a dive contribute to the nature of underwater decision-making. How someone responds in the midst of feeling bubble-brained determines much of their make as a diver. But their response is based also on their dive plan, training, and gear, or lack thereof.

Feeling bubble-brained emphasizes the importance of making decisions based on the plan, training, and gear available to that diver. Similarly, the trials of life exacerbate the importance of decisions based on reading the Bible, seeking godly wisdom, and living in community. Navigating the underwater landscape and the terrain of life are made easier when we engage in these practices.

Things may not go quite to plan, and the confusion of our own bubble-brain may contribute to some of our problems. To some extent, that is okay. What is important to remember is that when we are in the habit of relying on things, we are more likely to turn to them in a moment of turmoil, confusion, or being lost. We must practice what we desire to perform. As the heroine of *Jane Eyre* declared when faced with a life-altering and heart-wrenching decision: "I will hold to the principles received by me when I was sane, and not mad—as I am now. Laws and principles are not for times when there is no temptation: they are for such moments as this. . . . If at my individual convenience I might break them, what would be their worth? They have a worth—so I have always believed; and if I cannot believe it now, it is because I am insane."[2]

Which is to say, the Truth does not change even when our feelings do. If we decided something was right while we knew we were thinking clearly, then we can hold to it knowing that it is still right, even while our thinking feels muddled. If we decided on a plan while feeling safe in the light of day under a clear blue sky when the breathing was easy, we can know it is still a good plan when we become lost and scared in the darkened and turbid water while we breathe condensed air from a glorified can on our backs.

We practice in advance for the hard decisions that come later by living out our desired habits in the little mundane decisions of the day-to-day. This small decision-making matters because someday it will guide us in the big decisions. We must familiarize ourselves with God's Word, letting the Holy Spirit breathe on us, giving us a breath of his air and removing any symptoms of being spiritually bubble-brained. We can ask God for his superior wisdom, knowing that he promises to provide it to those who earnestly ask. And we can seek to place ourselves in positions of community, experiencing the challenges and joys of doing life together with others as opposed to "paddling our own canoes." When we learn to relish these relationships that honor and represent God's love, we may find that we are each propelled forward with impressive speed at the perfectly correct direction of God's leading. We can safely and effectively run the lost buddy procedure, navigating the hard decisions of life with courage and skill, even when we feel totally bubble-brained.

16

whale whispers

Once upon a time, in the tropical reaches of the southern Atlantic sea, lay the kingdom of Magic Castles. You could enter the watery kingdom only via map and compass, using listed bearings to follow a course over weedy strands of algae and baseball-sized nuggets of coral that littered patches of sand. *X* marked the spot where you'd find yourself entering the triangular-shaped kingdom ruled by the Magic Castles.

There were nine castles, rising from soft white sand and reaching upward through twenty-five feet of water to the surface. Streams of sunshine poured down on their towers, dappling their walls with moving light. The castles were alive, made entirely of coral. In fact, each castle was a single large coral. And each coral castle was special because they were *Dendrogyra cylindrus*, an endangered species of stony coral found in only a few parts of the world. The kingdom of Magic Castles was a particularly

special place because those nine individual castles were within swimming distance of one another.

Hovering before a towering castle, you would be spellbound. Thousands of miniscule, peach-colored, flowerlike tentacles extended from every surface, reaching out in soft, fuzzy, swirling motion. The flowers waved their petals like tiny hands in the water, simultaneously moving with and against the flow of water. If the coral castle felt uncomfortable or nervous, the tiny hands would curl into fists and withdraw into the peach-colored flesh, creating a thinly bumpy, velvety backdrop where before there had been thousands of blossoms. A few moments might pass before once again, the petals would relax and reopen, greeting the world around the castle, ruling the kingdom in a quiet and gentle manner.

Flurries of silver-striped fish rushed in streams around the castle towers like patrolling armies. Fluffy-headed Christmas tree worms poked in and out of the battlements at intervals as if they were busy maids feather-dusting internal quarters. Small crabs worked their way across the castles' soft surfaces, like servants waiting up and down hallways on any given errand. The wiry antenna of a crusty red lobster twitched as he stood guard at the base of a castle's dungeon. A green-blue grouper, fat-lipped and attended by two slender-striped wrasses, meandered her way down a colonnade, like a princess with her maids-in-waiting. But everyone in the kingdom knew that the real royalty were the coral castles themselves. Nine monarchs, rising from the sand of nothing, created a special place for every inhabitant of their magical kingdom.

For a long time, the location of the Magic Castles was kept secret. Only a few noble divers knew where it could be found. They served as knights protecting the kingdom, preserving it from being overrun by curious visitors who might not know or appreciate the value of the special living architectural inhabitants. Stories of the kingdom were myth. Tales of the knights were legend. And for many years, the castles lived in peace, quietly ruling their kingdom in the southern Atlantic sea.

But then came disease. The evil stony coral tissue loss disease (SCTLD) arrived with the violent fury of a vaporous dragon, infiltrating past even the most cautious guard, outperforming the knights by being untraceable until it had fully poisoned the whole of the royal family. The welcoming, waving hands of the peach-colored polyps withdrew to fists, then sloughed off in thick, mucus-like sheets of flesh. As the magic turned necrotic, the knight divers did everything they could to save the kingdom. For months, they fought repetitively at the site, applying a special potion they hoped would cure the monarchs of the terrible disease. But this potion did not work.

The soft, peach-colored blossoms sloughed away in the wash of current, leaving behind nine colonnades of bare white skeleton. The castles turned to stone, and their magic faded away. The kingdom crumbled.

Today if a noble knight ventures to the kingdom with map and compass, they will see only the skeletons of these ancient monarchs, standing as soulless reminders of what once was. But if you ask one of the knights who was there when the Magic Castles reigned alive and thriving, who fought against the vaporous dragon day in and day out, even while seeing the monarchs fall one by one until there was no one left to save and that noble knight hung her head in despair—if you ask her about the myth, she will tell you a greater Truth.

It might be silly to compare being a diver to the mythology of being a knight. But both are tough jobs that aren't always all they're cracked up to be. Don't get me wrong—diving is an incredible experience, and if you're able to give it a try, I highly recommend it. But it's helpful to know in advance that the staged photo shoots in scuba magazines and the curated photo albums of outdoorsy influencers don't capture the reality of diving in all its gritty splendor. Despite my best attempts to sincerely showcase the grossness of the occupation in my personal photos, there's just no visual way to capture the grime, ooze, and odors. It is helpful to know a little of what goes on behind the scenes so that the job of a diver isn't judged solely by the work they produce but also the effort that goes into the mundane aspects of their role.

Gearing up in dive equipment is in itself an entire process. By the time you finally wiggle your wetsuit over the top of your swimsuit and strap on your gear, you've transformed into a molten glob of sweat baking under the tropical sun. Splashing into eighty-something-degree water comes as a welcome relief, until your wetsuit begins to pinch at the wrists, elbows, knees, crotch, and neck. You tug on these places to fill your wetsuit with a thin layer of water, buffering the squeeze of neoprene against skin. The water coating your body comes with a layer of salt grit finer than sand, and if you have an itch, it's a challenge to scratch through the neoprene. After a lengthy dip in the briny sea, your skin gets fairly pickled, fingertips and toes pruney with soft-fleshed wrinkles. Because of the mask you wear to see, you can't inhale through your nose. And while breathing off a cylinder of compressed air via a regulator wedged in your mouth, the back of your throat becomes so dry that your tongue seems to crackle as it desperately twitches to salivate.

As a bonus, sometimes your equipment doesn't cooperate to give you a comfortable dive. Your mask might fog up, making it impossible to see, so you occasionally fill it with water to clear the condensation. Now, though, you can't see because the mask is full of water! By exhaling through your nose, water is forced out the sides as the mask fills with air. You can now see through the mask again. The downside? When clearing your mask of water, you repetitively blow your nose into it; so much for clean.

Sometimes a neatly coiled hose or line comes loose from your dive gear, snagging on a rock as you swim, requiring you to untangle and restow the item and be more aware of your safety. After a while, you start to get cold. Although the water around you might be 80 degrees, your internal temperature fights to maintain equilibrium at 98 degrees, causing your skin to prickle and inner core to shiver, as muscles occasionally cramp in your fin-kicking legs. As a sneaky strategy for briefly warming up, you find yourself relishing the relief of peeing in your wetsuit. The warmth spreads over your body like a comforting hug, before eventually dissipating in the coolness of the surrounding seawater. But probably the worst part of diving is what comes afterward.

When you resurface, you're excited to take off your mask and spit out your regulator. When you do, every orifice of your face begins to drip. Your mouth automatically drools, desperate for a sip of fresh water. Your eyes well up, blinking back salt and squinting against the abrasive sunshine. Your nose becomes a waterfall. All around you, resurfaced divers blow snot rockets. Mucus is everywhere, little globs floating on the surface and sticking to your gear. The only surfaces you have to wipe with are wet. Tired from the dive, you struggle back onto the boat, feeling the full weight of your heavy gear made heavier by its wetness. As you begin to remove gear, the smell of urine permeates the boat deck as unrinsed wetsuits turn rank in the sun. Welcome to the life of a diver!

Needless to say, diving is a unique sensory experience and not always comfortable, even to the most seasoned divers. In fact, it's sort of prerequisite that you have to be okay with moving far out of your comfort zone. Humans are not designed for the underwater world, so despite having technologically advanced dive gear and a high level of training and experience, you're still bound to encounter challenges and goofy, unpleasant moments. While this is true for any type of diving, one of the main distinctions between recreational and professional divers is that professionals have regular run-ins with the mundane.

When you dive frequently, even repetitively, the whole experience can lose its shine faster than the double-ended stainless steel clips you keep attached to your BCD. The first one hundred dives are a novelty, but when you're doing four or more dives a day, each of which is a "working" dive, everything starts to feel like a chore. Stepping onto the boat is like walking into a noisy office. Gearing up is similar to putting on a too-tight work uniform. The groupers who follow you on dives are annoying coworkers who constantly get in your way. You get frustrated when you're trying to tell your dive buddy something but they're not paying attention. With their back to you, they can't see you waving your arms to communicate in underwater sign language, and through the sound of their gurgling exhales, they might not hear you tapping on your diving cylinder to get their attention. You probably don't get much of a lunch break; between dives you find yourself scarfing down bites of limp mac 'n' cheese from a damp Tupperware turned lukewarm by the sun. It seems the only divers who eat fancy granola bars are the ones who dive for pure fun, because working divers are the kind to opt for cold beans. They bring a can opener and no spoon.

Depending on the kind of work you're doing, the dives themselves may be repetitive. You might make twenty dives at the same site, perhaps doing the same underwater procedure. You might spend a week diving in a coral nursery, doing nothing but scrubbing macroalgae off coral trees, which are human-engineered tree-like structures that hold growing coral fragments. As you rake off the algae with a massive metal brush, clumps of mustard-orange fire coral and translucent hydrozoans fall off and burn your skin with their nematocysts. Apply all the cortisone cream you want after the dive, and you might still be scratching the stings for days afterward.

As with any job, the thing that makes or breaks the workday is perspective. If you expect another sweaty, grueling day of physical labor in a dangerous environment, then you'll find yourself swimming in metaphorical hot water. But if you choose to shift your mindset away from fixating on discomfort, mucus, and hazards, then you might just find something amazing. No dive site best sums up this difference of perspective than Magic Castles.

Magic Castles, which had once been one of the best places to dive, rapidly transformed into a working diver's hell once SCTLD arrived. Most days, the dive team wielded massive caulking tubes of medicine, treating every sick coral by hand. We would work for over four hours underwater (a time frame requiring a series of dives) and treat only two corals, having to return day after day to treat all nine castles. It was a challenge of both mental determination and physical coordination to squeeze the handle of a caulk tube with one hand while pressing the stream of treatment down into the waving flowery polyps with the other hand, striving to cover every line of sporadic disease margin eating away at the living castles. Our hands would cramp up partway through dives, and we'd have to ambidextrously switch from using one to the other.

What was even more frustrating was that the treatment didn't seem to stick. The coral polyps were too fuzzy. When a strand of treatment was applied (much like smearing a line of toothpaste out of a tube onto a plush carpet), the flowerlike polyps retracted like fists. The treatment would stick until they decided to emerge again. Unfolding their petallike tentacles, they would slough off the treatment, rapidly undoing our work and creating patchy holes in the medicine. Sometimes it felt as if the castles themselves were fighting against our help. It made me wonder if these mortally sick disease patients knew from the start that they were in a losing fight and wanted to spend their last days in peace rather than battle.

Seeking advice from different agencies and researchers, we realized the desperation of our situation. Because the treatment was not sticking, death by disease was inevitable. But the idea of losing Magic Castles was unacceptable, so we got creative and crafty. We tried shoving bits of modeling clay into the crevasses of exposed skeleton, overlaying it on top of the medicine to stick it down. We tried stapling ribbons of burlap over the top of the medicine. But even when the medicine stuck, the treatment just didn't seem to work. And there was no better alternative.

For months, we divers, feeling anxious and frustrated, performed slight variations of the same actions over and over—always expecting, always hoping for different results. Despite all our hard work and endless trying, the magic of the castles slowly vanished. The death of this coral kingdom was heartbreaking beyond description. But defeat by death isn't the whole story.

I thought I was hearing voices. Amidst all the squeezing of the caulk guns, the scraping and smearing of medicine, the nibbling of butterfly fish, the crackling of shrimp, and the rhythmic cacophony of my own exhales, there were strange whisperings in the water. Sometimes there was a short-lived, high-pitched squeal, like the sound of my own ear canals equalizing under pressure. Absorbed and busy with stress-filled disease combat, I would hear the noises and voices and dismiss them as being sounds in my own head. It wasn't until we surfaced after a dive that the truth would come out.

"Did you see the dolphins?" my dive buddy asked.

"What? No!"

"A whole pod swam right in front of you and squeaked. There was even a baby with them. But you didn't notice. You didn't even look up!"

It turns out, the squealing was not in my own head, but rather in the water directly in front of me. I continued to hear noises underwater, and out of habit, I continued to ignore them. This happened many times. Absorbed in my work, I would ascend, only to be told by my dive buddy on the surface that we had been visited by a pod of dolphins yet again. Or a beautiful reef shark. Or a phenomenally large school of blue tangs. Whatever the amazing thing was, I had missed it completely.

The intervention work at Magic Castles was deeply discouraging, exacerbated by the additional disappointments of learning that I had yet again missed seeing some kind of incredible wildlife, frolicking right next to me. And so I finally decided to keep my eyes peeled. To not be so absorbed in the work in front of me that I missed the forest for the trees (or the corals for the sea, for that matter). I realized that despite all the frustration, discouragement, and disappointment of my current work situation, there was still beauty to be found. I didn't want to miss it anymore.

Not long after, I was treating a Magic Castles colony in frustration, discouraged by the fact that only half of its total surface area was covered in living tissue, the rest having been wiped away by violent disease despite being treated many times. Our medicine was not working. All our hard work felt like it was for nothing. And I was faced with the reality of watching the creature in front of me slowly die a terrible, suffering death.

Then I heard the whisperings around me in the water. The voices were very faint, partially musical, easily misconstrued with the raking shuffles of sand beneath my knees and the garbled bubbles rising from my regulator. Glancing around, I saw nothing. But the longer I worked, the more I thought I heard.

Remembering my determination to not be so absorbed in my work that I missed the world around me, I occasionally looked around from my vantage near the seafloor. Beyond the edge of the coral, the water was empty. There was nothing there, save sand, a few shells, and tufts of soft corals blowing in the current. I turned back to my work, acutely aware that the whisperings continued.

Eventually, I looked around again and was greeted by a strange sight. Partially camouflaged against the white sand, an opalescent fish about the size of my finger hovered in front of me. Standing on his tail, he gazed in my direction, watching with pupils outlined in neon blue shimmers on a golden-yellow face. I slowly started toward him, and the little creature sank tail-first into sand. Under him was a tiny burrow, a hole that was practically undetectable—the perfect-sized home and hidey-hole for this shy yellowhead jawfish. Now only his bright eyes and the tip of his head poked out of the hole, watching me.

I held still and waited. Cautiously at first, then all at once, the jawfish exited his burrow and swam forward a few feet along the sand, all the while making eye contact. And so I introduced myself. I explained the situation about the coral disease, which likely my new friend knew more about than I did (if only he could have talked!). He drew a little closer, darting his way in inches over the sand, the light glinting off his opalescent tail. I showed him the caulk gun and how to apply the medicine as he watched.

Across the way, my dive buddy gave me a curious look. I shrugged and pointed to the little fish, who immediately sped away to his burrow, backing in tail-first just as before. Patiently, I waited for my new friend to return. When he did, he brought another friend, likely his mate. There were, in fact, two yellowhead jawfish, each with their own burrow side by side. The burrows were so minute and undetectable that I had not seen either one until a fish emerged. The pair of yellowhead jawfish accompanied me for the rest of the dive, watching me treat their coral monarch while zipping back and forth between holes, alternatively observing me and flirting with each other in dancing circles over the sand.

"Another day, another dive," my fellow divers used to say before flipping over the side of the boat into the relatively refreshing water. As I treated corals, I continued to hear strange whisperings and have my doubts about whether these sounds were real or entirely in my head. Other divers had the sensation at times too, and although there was plenty of speculation, we focused on our work, sometimes delighted by the wonderful little things happening around us.

After first meeting the yellowhead jawfish pair, I looked forward to visiting them at their burrows each time we treated their neighboring magic castle. But they weren't the only ones I looked for.

One day I reached under the base of a coral, pulling myself sideways to get a better look at the underside of a ledge that needed disease treatment. From my sidelong view, cheek pressed into sand,

I could see into the darkened crevasse beneath the colony. Where I normally expected to find lobsters or crabs, I discovered a far more sinister creature lurking. I cautiously pulled my hand away from its grip on the coral's skeleton. Only a couple of inches away perched a spotted scorpionfish.

Camouflaged to blend in with a variety of red, green, and brown algae-encrusted rock, the scorpionfish glared at me with mottled eyes. I could see its gills gently opening and closing, as its venomous dorsal fin barb gently pulsed with each breath. I backed away, thankful that I had not placed my hand two inches farther to the side. If I had been stung by the venomous barb, I would have experienced excruciating pain for twelve hours, followed by other side effects like skin necrosis, vomiting, and tachycardia. Some fish are better left alone.

Ironically, the creature that gave me the biggest scare underwater was relatively harmless. It had been a long hour and a half dive. My leg muscles were cramping from overuse, while my hand ached from squeezing the caulk gun and my fingertips were cut from pressing medicine into the sharp grooves of coral skeleton along the disease margin. I was daydreaming of sipping sweet iced tea, sitting in air-conditioning, and blowing my nose into a dry Kleenex. Suddenly, a large green monster was cruising through the water toward my shoulder. I turned, receiving perhaps the biggest startle of my life.

The creature was rapidly gliding toward me with intentional arm beats. Her skin was forest green leather. Her shell was as large and thick as a coffee table, overgrown with a soft layer of green hair algae. Her large, dark eyes took me in, seeming strangely human with their wide-eyed gaze of wonder. She circled toward me once, nodding her head the way a fair maiden might acknowledge a knight, before turning and vanishing into the misty blue abyss.

After seeing the loggerhead sea turtle, I began to understand the deeper thing that was happening in the backdrop of my dives, even without me knowing it. When I went about my work as usual and expected to experience nothing else, my day was filled with the mundane. My limbs ached, the work was toilsome, and diving was a chore. But when I entered the water determined to keep my eyes peeled, to see the underwater world for all its God-created splendor, then my day was filled with beauty. I still ached, the work was still laborious, and diving was still a chore, but instead of focusing on only those things, I was able to enjoy what God had created in that space. I was able to see his good work overlaying and transcending my own.

I was no longer entirely absorbed in thoughts of the corals' imminent demise. I was able to enjoy the environment that flourished around them, even as their kingdom slowly fell. In this way, I was able to have a new sort of faith. "Now faith is confidence in what we hope for and assurance about what we do not see. This is what the ancients were commended for" (Hebrews 11:1-2, NIV). Because I changed my perspective from focusing only on my work to observing the work God was already doing, I was able to have certainty in his ability, rather than continually recognizing the inadequacy of my own. I was able to have certainty in God's Kingdom, resting in the eternal assurance of his promises, instead of stressing about the failure of this kingdom on earth. I was able to be sure of what I hoped for—that the story was not over for these corals, that the kingdom they ruled and the work I was doing mattered even if not through the lens of human eyes.

I was able to be certain of what I could not see—that even as the corals died, God had given all of creation a promise that someday he will make "everything new!" (Revelation 21:5). Faith enabled my understanding of God's Kingdom to transcend my immersion in and obsession with the earthly kingdom. God taught me to understand my work through the perspective of eternal impact for his Kingdom rather than through the mere failure of my earthly endeavors.

I did figure out the voices. It took a while to discover, but the nine *Dendrogyra cylindrus* colonies—the Magic Castles themselves—were singing. The thousands of flowerlike polyps had been raising their thousands of voices like a symphony and choir of otherworldly praise. There they were, praising their Creator before me for all their days. And because of my distractibility and wrong perspective, I had almost missed it. What a tragedy it was the day their voices went silent, muted until the future arrival of the completed eternal Kingdom.

During one of my last intervention dives at Magic Castles, the voices of the coral polyps reached their greatest crescendo. The corals were singing their hearts out, even as they died! And this time, they were accompanied by a chorus of louder voices.

I recognized the voice of the corals, but there was another voice—melodious, resonant, and unfamiliar. I thought it might be in my head, until my dive buddy swam over excitedly, holding up an underwater clipboard on which was scrawled the question, "Do you hear the whales?"

The whales sang a multiharmony duet with the corals. The great blue voices of the faraway deep intermingled with the softer symphony of floral melody, as the peach-colored polyps blossomed, bent to the pull of the current, and succumbed to their end. I savored their song, and I remember it still—the choir of coral and the whispering whales.

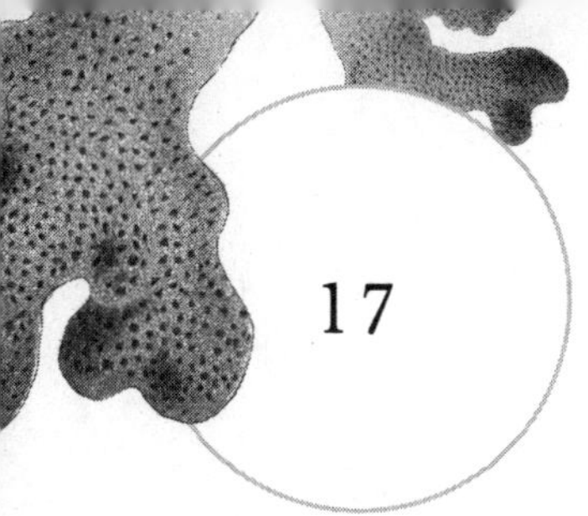

17

silhouettes in the maze

Life can feel like diving in the Maze, a site full of shadows. Murky blue-gray abysses lie beyond the cliff edges, swirling through sheer canyons, darkening the underside of every ledge and crack. The twisting walls and cliffs of the Maze are filled with caves and swim-through tunnels. Alien-like life dangles from the walls in the form of leathery algae and encrusting sponges. Where corals grow, the landscape is angular and sharp, with punctuated bursts of color lying in shade, spreading wide in rarer patches of shallow sunlight. Flurries of fish fly between ridges like barn swallows darting aggressively between buildings and tree branches. In rocky tunnels and caves, bigger fish await. Flabby-lipped goliath groupers the size of cars snooze within their territory's shadows. Razor-toothed barracuda hover in the canyons. Silver-silhouetted reef sharks patrol the Maze's perimeter. For all its juxtaposition of

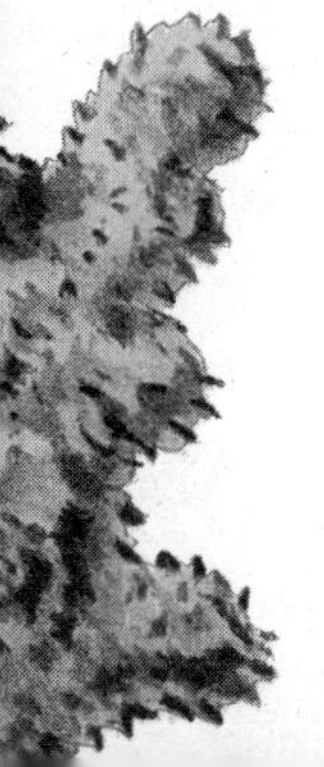

dark and light with large and unexpected creatures emerging and withdrawing within its labyrinth, the Maze is frequently described as creepy. And yet it remains my favorite dive site.

The eerie feeling that eats at the edge of my mind feels more like an invitation than a threat. The cliff edges are distinct markers to the unknown because from atop the sunlit ridge of the reef, you cannot see what lies within the depths of its labyrinth. I cannot help but wonder what is there. And so on a disease reconnaissance dive, many a dive buddy has watched me descend quietly over the cusp of the edge into the cloudy blue-gray abyss, with only a thin stream of exhaled bubbles to mark my path as my blue fins vanish from sight.

They cannot see me over the edge. But it is over the edge that I can see everything. My pupils dilate in the darkness, and the grayish bits of fuzzy and globular shapes come into colorful focus. Vivid orange cup-shaped sponges gleam beside the pastel *Porites* colonies' wispy polyps, their ever-moving tentacles combing for particulates like thousands of miniscule hands waving hello. A family of dinner plate–sized angelfish float past, rainbow-scaled with neon blue chins and yellow-tipped fins. A stoic stone crab watches my eyes pinch up into a smile at the sight of his funny little mouth, gnawing on some scavenged bit held within the pincers of his thick burgundy claw. Between walls of this labyrinthian home, I float weightlessly to the sandy bottom, white and lightly confettied with shells. A pancake-like figure, brown as chocolate, glides over the sand with happy wingbeats, turning ever so slightly to the left and then right in a stingray's dance. A feisty little damselfish emerges from a cleft in the wall to nip at its own reflection in my dive mask, whacking its body against my face in a series of eager kisses, intended to startle me, until I shoo it away as if batting at a fly.

From here in slanting shadows, I gaze upward, where streams of light beam through the ever-moving sky that is the surface, filtering through water to rest on ridgetops. Crisp shapes of corals, algae, fish, and sponges line the edges, perfectly framed in light. The silhouette of my dive buddy moves among them. As I float in the depths where all the distant lines are sharp and clear and every shade of light is beautiful, I can see the diver's silhouette moving uncertainly through the maze of life, wondering at the shadows. It is with the same measure of clarity that, in the twenty-twenty vision of hindsight, I can look back on my journey and see that God was there in the maze of life with me all along.

I don't always feel this way. There have been numerous seasons throughout my life when the answer to my prayers is silence, the revelation while reading the Bible is stillness, the counsel of trusted friends doesn't break the darkness, and the response of my heart is despair. In these seasons, I cry out wondering where God went because it feels as if he has up and left me. I feel alone, stuck here, doing all the right things and trying to engage. Despite my efforts to see him and hear Truth, I keep coming up empty. It is as if my prayers bounce off the ceiling and the Word of God returns void. Have I gotten everything wrong? Is this what relationship with God really looks like? I've heard it said that "the only person who dares wake up a king at 3:00 AM for a glass of water is a child. We have that kind of access."[1] But sometimes my need feels far greater than a glass of water. I feel like the desperate child banging on the door of the bathroom, begging the haggard parent to notice me when they're just trying to do their business. What if I've gotten everything wrong about God?

But the Bible's story demonstrates that God is never haggard and that you and I are not mere nagging inconveniences. Rather, each of us is called his precious child. These truths do not change when our feelings do. (If they did, they wouldn't be Truth.) But my goodness, feelings do change and life is hard! Our circumstances change all the time. Life is unpredictable. Things don't always turn out as we planned: "We can make our plans, but the LORD determines our steps" (Proverbs 16:9).

Faced with disappointed expectations and unmet hopes, it is easy to slip into a pit of despair. When the voice of the one said to "determine our steps" is silent, it is far easier to remain in the pit than it is to navigate out of its miry grasp. There in the depths, we tend to lose sight of the things that were once vibrantly clear to us.

In a season of silence or despair, we may find solace in the Bible. The Psalms are far and away the best biblical portrayal of this pendulum swing of feelings and the trials of life. A collection of poems and songs, the majority of this book was penned by King David, a man whose life and spiritual walk were varied and volatile. Beginning as a baby-brother shepherd boy who was anointed the future king, David killed a terrifying giant in a battle against a nation that rejected God. Later, he was forced to flee for his life as he was hunted by his own nation's current king, whom he was destined to replace. As he moved from cave to cave and forged a band of mighty men, he made music and penned poems. David eventually gained the kingship, lived in the palace, and ruled the nation, coming to be known as a "man after [God's] own heart" (1 Samuel 13:14). And yet he committed one of the most notorious acts of adultery in all of history, leading to a man's murder and the death of his own infant son. Later, one of his adult sons staged a coup against him and did numerous other awful things. From an outside perspective, the life of David is interesting. But from the Psalms he wrote, we get an insider's taste of the turmoil that rocked his world.

While his circumstances may be drastically different from our own, the emotions David expressed in his writing are relatable. In my experience, reading David's words during seasons of spiritual silence has been profoundly cathartic. Even as I feel alone and left behind by God, it is comforting to hear the words of someone else who many times felt the same way. And yet we know that David was not left by God. His history as king and the ebb and flow of his writing reveal that God did not leave or forsake him. For these reasons, the Psalms are both a comfort and reassurance to the confused heart.

Another compelling biblical portrayal of the trials of life is the book of Job, which explores the concept of whether a man will wholeheartedly serve God even while facing profound suffering and receiving nothing good in return for his devotion. Job's response in the face of suffering evolved over the course of time, and his dialogue is a fascinating study, particularly for anyone who has felt neglected by God. Notably, nearly the whole book of Job is dialogue, save a brief introduction and conclusion. And of that dialogue, most of it belongs to Job and the friends who accompanied him through his grief.

God did not speak to Job until near the end. Effectively, he was silent throughout the majority of the book and thus he was silent throughout the majority of Job's season of suffering. Like Job, we may wonder why. We must read the whole story, including God's words, to find out. But because of Job's brutal honesty, the book offers practical insight for those of us who struggle with the feeling of God being silent. Job struggled too. To say he felt that God's silence added insult to injury would be an understatement. Job cried out to God, questioned him, and eventually demanded justification. God's response was that of an all-powerful Creator who knew Job deeply, who was present throughout every moment of the situation, and who offered blessing in the end.

Biblical examples are well and good, but they are helpful only if they give us practical tools and meaningful implications. The Psalms are highly relatable and demonstrate the back-and-forth swinging of a man's heart, rent before God, as David sprinkled in statements of Truth amidst the more volatile emotions and circumstances. Job portrays how a righteous man may act in the face of profound suffering and what sort of response to our questions we can expect from God. But comfort and hope take us only so far. How do we actually go about living with these concepts? With the full scale of emotions available to us, how are we to navigate their volatility and the seeming changes to Truth they incur? It is right and honest that we wrestle with these questions. Both David and Job did, and there will be seasons when it is our turn.

As my dive buddy watches me disappear over an edge of the Maze, I feel as though I am slowly free-falling into a dark canyon. The blackness below seems to reach up to swallow me, gently revealing its creepy inhabitants. Shadows wrap around me. In this deep crevasse between textured walls of rock and coral, overgrown with algae and crawling things, I hover over a soft patch of creamy sand and take stock of my surroundings. Bubbles gurgle from my regulator, spluttering a stream of glistening white orbs that vanish upward. As my eyes adjust, little treasures come into focus. A neon yellow splotch of porous sponge juts out from one wall as a thin, orange speckled goby settles atop the swirls of a buttery brain coral. Spiraling green tendrils of soft coral fronds waver in the current. A mated pair of gray French angelfish flirt back and forth on their journey through the canyon. A wan-faced grouper accompanied by three silver, suctioning remoras glides effortlessly past.

I turn to look back over my shoulder and glance upward to where sharp outlines of corals, algae, and other photosynthetic-reliant organisms fringe the canyon's edge, spangled in sunlight so bright that I squint. The meandering, darting, and dancing silhouettes of fish punctuate the scene above, at the center of which is my dive buddy, a sleek presence softly floating, working her way along the edge, all the while looking for me.

I am never far from my dive buddy. But the fact that she cannot see me drives home a realization. I can see her, and I am assured of her presence as I watch her silhouette glide above the reef. And yet I am invisible to her sight. She watched me disappear into the blue-gray abyss, a trail of bubbles to mark my path, but she can no longer see where I am. My dive buddy knows that I am near, but since she can no longer even see my bubbles, she must rely on the knowledge of my character and our relationship to know that I am not far from her. That I have not left her. That even though she cannot see me, I am keeping a keen eye on her.

Perhaps it is like this with God. God can see in all places, even the darkest and most shadowy ones. Anywhere we go, most assuredly, he is there first. If we are in a pit of despair, he is with us, even as he is silent. We may not be able to sense him, but as we seek to accomplish his will, he promises to go with us.

> But now, O Jacob, listen to the LORD who created you.
>   O Israel, the one who formed you says,
> "Do not be afraid, for I have ransomed you.
>   I have called you by name; you are mine.

When you go through deep waters,
    I will be with you.
When you go through rivers of difficulty,
    you will not drown.
When you walk through the fire of oppression,
    you will not be burned up;
    the flames will not consume you.
For I am the Lord, your God,
    the Holy One of Israel, your Savior."

ISAIAH 43:1-3

We may feel alone, but God's presence is there beside us, perhaps just out of sight. When we are required to dive into unknown waters, we are told that God is there too. When we are called to swim through deep rivers, we know that God goes with us. When we must walk through fire, we remember the one who is greater than the flames. When we are humiliated or frustrated or anxious or talked down to, or even when we are called to do something that is impossible, we can be assured that God goes with us and that he is active. When God is silent as we cry out to him, we can know that his quietness has a purpose. If he chooses not to speak, it is for a reason. We may not be able to see or hear him now, but that does not change the Truth that he is with us and that he sees and hears us. His presence and his love have not left us. He promises they never will.

The silhouette of my dive buddy reminds me of another aspect of God's presence. As the diver shrouded in darkness at the bottom of a marine trench, I have to look over my shoulder and up in order to see her. Her presence is visible to me, but sometimes only in hindsight. The same has been true of God's presence in my life. Only after coming out of seasons of silence and the pit of despair did I eventually see that God was with me all along. It took a while to see Truth.

Truth does not change when our circumstances do. It remains constant when everything else in our world is volatile, when we are plagued with uncertainty, doubt, and despair. When we choose to seek after and remember Truth, we are able to wait for God's presence.

> I waited patiently for the LORD to help me,
> and he turned to me and heard my cry.
> He lifted me out of the pit of despair,
> out of the mud and the mire.
> He set my feet on solid ground
> and steadied me as I walked along. . . .
> Oh, the joys of those who trust the LORD.
>
> PSALM 40:1-2, 4

I do not want to minimize the tremendous challenge that seasons of spiritual silence bring. But we have the assurance, both from biblical Truth and our life experiences, that God will never leave nor forsake us despite circumstances that make it temporarily impossible for us to see and hear him. In the darkness of the Maze of life, it's helpful to remember that the presence of the one there with us is sometimes seen only through the perspective of hindsight. What is unclear in the moment may be revealed at a later time. God was with us in the fire, river, and flood all along. We may have felt as though we cried out to nothingness, but he heard our cry and was there too.

In these hard moments, beauty still abounds, like being surrounded by unique creatures and sunlit silhouettes in an underwater canyon. Dark times may offer the greatest potential for growth, beauty, relationship, and interaction with the Kingdom of God. The silence is where we learn to listen for his voice. The darkness is where we see the streams of light pour in, making the true shapes of things clear. The stillness is where we find space to sit still, to seek Truth and relationship with God in raw and unique ways. Despair can be a posture in which we experience his healing and rescue as we're drawn into greater reliance on him. Not only may there be a purpose to these seasons, but maybe they are designed by a loving God to provide us with immeasurable blessings.

Each time I dive into the Maze, the familiar eerie feeling ebbs at my mind once again, inviting me over the cliff edge into a hazy, blue unknown. In some sense, of course, I already know what's down there. As I descend into shadows, I gaze upward to see light pouring itself onto the colorful ridgetop, edged with life and glory, as the distinct silhouette of my dive buddy glides smoothly across the ridge. In the silence of the depths, I feel as though I am looking back on a moment in time, a place I have been, with a clear understanding of God's presence—as though I can see his silhouette in the Maze.

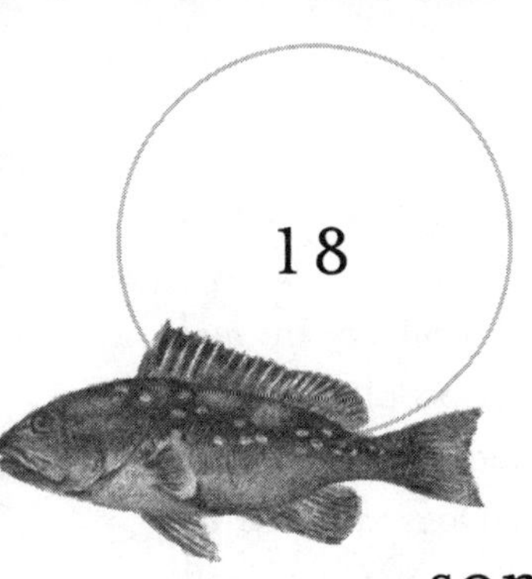

18

someone is watching

My first pets were a pair of goldfish, Goldy and Silvery. I was utterly devoted, watching them for hours at a time. I loved feeding them and seeing how excited they were gobbling up the little flakes I sprinkled across the water's surface. It was my favorite part of *Mister Rogers' Neighborhood* too—watching the fish dart to the surface whenever Fred Rogers scattered food in their aquarium. I noticed how happy Goldy and Silvery were after Dad cleaned their tank. They seemed to frolic in the water. When it was winter, they swam slower and slept a little more often, resting their bellies on the rocks. During summer, they swam faster. On especially hot days, they stayed near the bottom of the tank, seeming to pant until I dropped a few ice cubes in the water. They'd swim up to nibble on the edges of the cubes, relaxed and comfortable again at the proper temperature. After several years, we upgraded their tank to a larger one. With twice as much room as before, my pets rapidly doubled in size.

I spent so much time watching Goldy and Silvery through the years that I learned their personalities. Some people will think that's a bit of wild thinking—that fish could have individual personalities—but then, why wouldn't they? Dogs and cats do. Why wouldn't fish?

Silvery was slightly bigger and moved more methodically than Goldy. He took his time browsing from one end of the tank to the other, thoroughly enjoying sucking up rocks at the bottom of the tank before spitting them back out again. Goldy was wigglier and spry. He breathed a little faster and often beat Silvery to the food by a couple of seconds. He liked to follow Silvery like a shadow, and after Silvery died, Goldy moved slower and was never quite as enthusiastic about feeding time. Those fish were with me for my whole early childhood, dying at ages ten and eleven, respectively. But for all the time I watched my fish, I don't think it ever occurred to me that they were watching me back.

Someone once suggested to me that my parents had "replaced" Goldy and Silvery multiple times throughout my childhood, since goldfish "just don't live that long." That person was wrong for three reasons. First, if Goldy or Silvery had been replaced, I would have known right away because I knew my fish. Second, I asked my parents directly. No, they had never replaced my fish. And given the opportunity, they would not have on principle. They believed in the importance of teaching their kids about the reality of death, opting to help us grieve and process rather than keep us in ignorance and defer inevitable pain. They taught me to love hard and well, even as death comes—and that lesson began with the deaths of my fish.

Lastly, anyone who says that goldfish do not live to be a decade old has clearly not done their research. With proper care, goldfish can live between ten and fifteen years. Some varieties can live as many as thirty years.[1] The age-old story of a new pet goldfish dying after a day or two is a tragedy and often reflects a lack of knowledge and care on the part of the pet owner. That level of profound negligence for an animal's life would never be tolerated if applied to cats or dogs. And yet the pet industry does not track this; neither does society condemn those who participate in the mistreatment and abuse of fish.

I'm with Abraham Lincoln, who was quoted as saying, "I care not for a man's religion whose dog and cat are not the better for it." How you act toward the least reflects your beliefs toward the greatest. If the treatment of dogs and cats reflects religion, then the treatment of fish does as well. But I digress.

I cannot unsee how much sea creatures are like humans. There are introverts and extroverts. Some live solo or as a pair in a private home—maybe in a branching coral colony, tender sea anemone, or elaborate underground burrow. Some fish prefer to travel in large groups, schooling to travel miles through the ocean. Most creatures appreciate a good meal and have specific preferences for what and how they eat. There are the vegetarian rabbitfish and surgeonfish, the meat-loving frogfish and barracuda, and those willing to eat everyone else's leftovers, like remoras and crabs. There are the neat freaks, like cleaner shrimp and sand sifting sea stars, constantly patrolling their area to vacuum every undesired speck they find. There are the romantic ones who try really hard to get the girl, like pufferfish, who put on elaborate, decorative displays to attract the attention of potential mates. There are the intelligent, highly self-aware fish, like wrasses, who can recognize their own reflections. Some fish struggle with anxiety, like tangs, whose vibrant blue-yellow colors can instantly fade to a dull gray when they are stressed or feel threatened. And just as with people, some fish, without necessarily meaning to, can hurt you.

I have been barbed a handful of times by sharp dorsal fins while trying to get ahold of a defensive fish. It's a hazard of the job. Working in aquariums, I learned that one of the gentlest methods for catching a small fish with the least amount of stress to them is to catch it bare-handed. Nets are great for convenience, but they're rough on fish. The mesh netting can stick to scales and strip protective mucus from the fish's body. Wet skin is much softer. But this is a pro move. To be successful, you have to know what you're doing. You calmly swoop your hand into the tank, slowly back the fish near a wall, and rapidly scoop, positioning your thumb directly over the dorsal fin and the rest of your fingers flush with its belly. If you do it right, the fish is caught before it even knows what you are trying to do. If you do it wrong, you get a frightened fish in hand with your thumb full of needle-sharp barb, its painful toxin slowly pulsing into you. It usually feels like a beesting. I've often wondered what it feels like to the fish.

Although I've gotten a few barbs from tangs, I have been bitten by a fish only once. It's strange to think that for all the hours I've spent watching fish, it wasn't until one bit me that I gave much thought to them watching me.

I was immersed in my work collecting genotype samples for a genetic disease study. This process required me to sample twenty target corals per dive by breaking off a tiny portion of colonies with the back tip of a mason's hammer. I'd catch the sample as it fell and place it in a labeled ziplock bag. Then I would hammer a nail with a cattle ear tag into the substrate beside the coral, place a measuring stick next to the colony and take several photos, make a few notations on a clipboard data sheet, and move on to the next coral. Once I found a rhythm in the process, I could sample at a rate of one coral every two minutes, which meant I could finish sampling for a dive just inside of forty minutes if I was concentrating well and did not get interrupted. I liked to finish as quickly as possible because it allowed time to leisurely enjoy the dive while my dive buddy wrapped up his sampling.

I was cruising along with my samples that day, intent on the movement of my hands and the next step in the sampling process. I had just finished carving out a sample fragment and was placing it in a plastic bag when a red grouper suddenly darted out from the corner of my peripheral vision. He snatched the bag with his teeth, shook out the sample, dropped the bag, and then floated motionless while staring at me.

The event was over in a flash. I was stunned. Then frustrated. He must have liked the shininess of the light reflecting off the folds of the bag and decided to figure out what it was. I searched for the sample, but the tiny piece was lost among a maze of crevasses and moving fronds of thick algae. The grouper had just undone my last bit of work, and now the coral before us would pay the price of providing another sample.

Fish are rather like people in how they exhibit curiosity. But they don't speak human languages to ask questions; they stare. And they don't have arms to shrug their shoulders with, so they do body tilts, angling themselves slightly in the water as if cocking their heads with their whole selves. And they don't have fingers to poke things with; they have mouths.

Some species of fish exhibit more curiosity than others, but I think groupers are among the most curious of all. With big bodies, thick heads, and enormous droopy lips, they seem like the overweight couch potatoes of the underwater world. But their behavior gives them away. They can be territorial, like dogs ready to defend their backyard from unwanted intruders. Groupers can bark. Often described as a boom, groupers can emit a vocalization much like that of a bass drum, shouting vibrations through the water that can be both heard and felt. The vibration of their barks, simultaneously amazing and terrifying, can rattle a diver's nerves.

Endangered goliath groupers can look like cars underwater, being over eight feet long.[2] But in the southern United States and Caribbean, the groupers most often sighted are relatively small. Reds and blacks tend to reach one to three feet in length. They love to follow divers, ambling slowly behind or perching alongside your peripheral vision while swimming, silent and always watching with their big, brassy eyes. As a diver, being followed became such a common experience that I tended to ignore groupers after the first couple of minutes of each dive.

Now, however, I watched the grouper warily out of the corner of my eye while I resampled, trying not to whack my thumb with the hammer. Sample finally in hand, I placed it in the bag. The grouper gazed, unblinking, as fish do. He did not move. I half expected him to bark at me, but he did not. I sealed the bag. The grouper repetitively opened and closed its mouth, casually showing off his fine, razor-sharp rows of teeth, as if smiling. Irritated, I quit paying him attention, wrapped up my work at that coral, and swam to the next.

I didn't even get the chance to sample this coral. I had forgotten about my grouper companion until the second fall of my hammer.

My left hand was braced against the substrate for balance against the blows of my right hammering hand, which also held a ziplock sampling bag tucked between my fingers.

I blinked just in time to see the grouper fly forward. He bit into my hand while trying to rip at the bag, and received a generous whack across the face in the process from my pull-back momentum. Startled at the sudden movement and pain, I dropped everything, thankful to be wearing work gloves. The mischievous fish was chewing on the bag, looking as irritated as I felt. Suddenly, he dropped the bag and attacked the hammer in a frenzy. The metal tool would not be punished, and when the grouper realized that it would neither swim away nor retaliate, he tired of it.

Then it was the grouper's turn for discipline. I gave him a bubble-filled lecture on the importance of the work I was doing and how inconvenient his interruptions were. I then instructed him to back off. Undoubtedly curious about the sounds this funny creature was making, he drifted nearer to my head. I could see his eyes slowly registering the reflections of light in my dive mask. His curiosity was mustering bravery again, and I realized that I was about to get a face full of grouper teeth. I had to do something, fast.

And so I spanked the grouper. My arm whirled out and planted a hard underwater pat on his side. Instantly, survival instincts usurped his stubborn curiosity. He was gone in a flash, hiding some forty feet yonder under a ledge. But still watching.

I finished the dive dubiously, frequently glancing over my shoulder and checking the blind spots beside my elbows as I worked. Nobody there. But I was jumpy and careful. And when I finished sampling, with a mesh bag full of ziplocked samples, I could spot the grouper gazing upward, watching me ascend to the glittering surface and out of view.

As at the end of every workday, I returned to the team house where my coworkers and I lived together while in the field. Hair still wet from my shower, I was reading in my room, sideways on my bed with my back against the wall, when there was a knock on the doorframe. One of my coworkers peeked in. "Hey, I know it's late. But I forgot to make dinner earlier. Can I . . . ?"

"Of course."

He walked through my room to the kitchen.

My room adjoined the kitchen and living area. It was less of a private room and more of a hallway or office. With only two private bedrooms available in the house, I had opted for the unwanted downstairs room and given my team the rooms upstairs. I didn't mind the frequent evening interruptions, and having the nicer rooms made them happy. I nestled back into my pillow and kept reading. Savory smells wafted from the kitchen.

"Have you eaten yet?" my coworker hollered.

"Yeah."

"What was on the menu tonight?"

"Mac 'n' cheese."

"Classic!"

There was a significant pause, long enough for me to continue reading. I could hear oil sizzling across a pan and a spatula stirring. Suddenly, my coworker was back in my bedroom, standing in the middle of the floor, telling me how the air compressor we used to fill our diving cylinders had been giving fits lately. Then he was talking about the hot sauce his aunt had given him for Christmas. And the jump in conversation did not seem strange.

He went back to the kitchen for a few minutes and returned with a massive bowl of hot-sauced stir-fry. I glanced up from my book again, expecting him to pass through the room and head toward the living room, where he normally ate. He lingered by the doorframe.

"What are you reading?"

"Psalms. From the Bible."

"You read the Bible every night, don't you?"

"I do. It's important to me."

He nodded, looking down at his food, then back up again.

The significance of the moment hit me slowly. We'd been working together as vague acquaintances for over a year. My early attempts at knowing my teammates had been firmly rebuffed. And so every night we made dinner in the same kitchen and ate in the same living room, but we stayed out of each other's lives. In my open hallway of a bedroom, filled with noise and smells and interruptions, my coworkers had seen me reading. And now one was hovering by the door instead of passing through to the other room.

I slowly closed my Bible and set it aside. "Hey, do you want to sit down?" I hesitantly gestured to the chair in the corner across from my bed.

"Sure! I'm in the mood for a bit of a chatty-chat."

I laughed. And we chatted. We talked about psalms, corals, work, our jobs before this one, favorite foods, and our families. We laughed so loud that we didn't notice another coworker's head poke around the doorframe.

"Whatcha guys doing?"

"Oh, just hanging out. Wanna join?"

"Sure!" He casually walked to the bed and hopped up next to me. "What are we talking about?"

I fell asleep significantly later than usual that night. But before I did, I lay in my sleeping bag watching the spiders dance across the ceiling and praying for my coworkers. That's when I remembered the grouper.

There seemed to be a connection between the grouper's curiosity toward the work I was doing and my coworker's curiosity toward my reading. Both had watched me quietly for a time. Both had observed the consistency of what I was doing. And both had unexpectedly interrupted that. Of course, my response had varied dramatically in these situations, but one thing was clear from the start: I hadn't known they were watching.

We may never know who is observing us. For those who follow Jesus, the love we have for him should be evident by our actions, which are witnessed by the people around us. These witnesses may or may not choose to ask the reason for the hope we have. But when they do ask, we need to answer. As the apostle Peter admonishes, "If someone asks about your hope as a believer, always be ready to explain it. But do this in a gentle and respectful way. Keep your conscience clear. Then if people speak against you, they will be ashamed when they see what a good life you live because you belong to Christ" (1 Peter 3:15-16). Whatever the interruption, even if pain or distraction or irritation, we must be prepared to explain why we live the way we do, how our actions connect to our love for God and are made possible by his great love for us.

"Why are you reading Psalms?" was a straightforward question. It was also a blatant opportunity to lovingly share the transformative Truth of what it means to have a personal relationship with God. Praise him, I was prepared to answer the question when asked. My coworker learned not only about Psalms, but about my relationship with the Creator of everything we had seen while diving that day. My words built a spiritual window to the terrible beauty, fearless hope, and relentless restoration that overlays the world as a marine biologist sees it. My coworker stood at the window that night and gazed.

Granted, not everyone responds the same way. Some people ask careful questions as they eat dinner, while others aggressively attack or bite. The apostle Peter, who faced much persecution throughout his life, writes encouragement for those times when people respond negatively toward you because of Christ. He explains that even a bad response is a sure sign that God is working in and through your life. If people hate the Jesus in you, it's because they see him in you. If they reject you because of him, they had to first identify him in you. In an unworldly and ironic way, it's a sort of compliment. The Bible calls it a blessing: "If you are insulted because you bear the name of Christ, you will be blessed, for the glorious Spirit of God rests upon you. . . . It is no shame to suffer for being a Christian. Praise God for the privilege of being called by his name!" (1 Peter 4:14, 16).

Jesus himself provides a clear picture of how you should understand and respond in such a situation. "Blessed are you when people insult you, persecute you and falsely say all kinds of evil against you because of me. Rejoice and be glad, because great is your reward in heaven, for in the same way they persecuted the prophets who were before you. . . . You are the light of the world. . . . Let your light shine before others, that they may see your good deeds and glorify your Father in heaven" (Matthew 5:11-12, 14, 16, NIV). You are blessed. Even in the face of insults and persecution, you should rejoice and be glad—not in an insincere or fake way, but out of understanding that God sees and knows. That you are being treated as he was treated. That perhaps you are being made to look more like him.

It is significant that these verses use the wording "*when* people insult you" rather than "*if* people insult you." The Bible clearly states that Christians will face hard things as the result of Jesus being displayed through their lives. But far more than a mere consolation, these verses in Matthew are a call to shine light brightly in the world, even as you encounter hard treatment from people. No need to spank the metaphorical grouper; leave judgement to the ultimate Judge and remember the promised blessing. Keep doing the labor God has placed before you, be it genotype sampling or reading your Bible in bed. Remember that someone is watching.

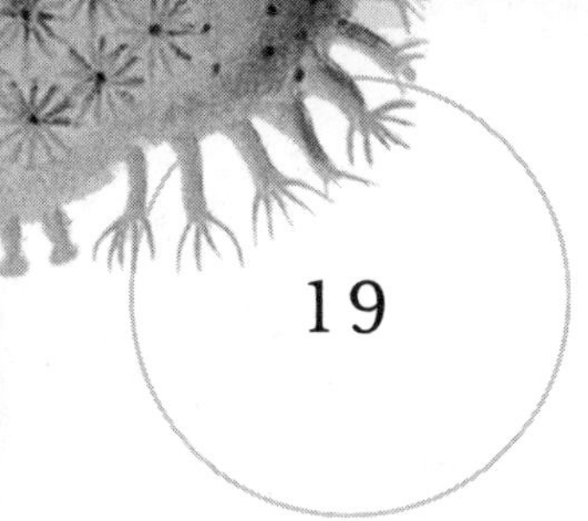

19

tending wild gardens

We were sitting in the back patio garden, surrounded by a thicket of carefully potted plants. Among them were a spritely coffee bush, bundles of pepper sprouts, a tender lime tree, and a ten-foot-tall Peruvian apple cactus. In the background, the low buzz of static crept from a radio inside the house. Mosquitoes hummed shrilly as chirping birds darted above the branches overhead. The solar-powered dragonfly bulbs gleamed charmingly in the fading light. Leaning back in tattered lawn chairs, boots off and dirty socks tossed aside, we sat. Sometimes in silence, sometimes in stories. This garden was his private, creative space. Because of him, it was my solace.

It seemed an unusual friendship. It struck park staff as odd that the incredibly tall, sometimes crotchety, and notoriously quiet head of maintenance and the young, blonde, friendly coral team

leader could get along so well. We were perfect foils of each other. But in an environment of noise and restlessness, I was drawn to his quiet, sensing there was wisdom behind when and what he chose to speak, as indeed there was. He liked that I listened. Given time sitting silently in each other's company, he let me earn his thoughts. He told me once, "The garden wall has always had an opening. I never made a door. Nobody who came into the garden stayed, until that first night you came." Quiet and solitary though he was, he told me stories about his forty-one years in the National Park Service, the old diving days, his childhood in Pennsylvania, his daughter who lives in Seattle, his Catholic upbringing, why he liked mowing the lawn, how to make a proper cheesesteak, and his name.

He was called Tree. As bizarre as it was, that was the name listed on his credit card statements. (I know because he showed me.) Always the skeptical scientist, I asked him once if that was indeed his real name. His bright blue eyes flickered that alive, knowing look, and he whispered the truth. I was sworn to secrecy.

In the dim light of dusk, silent tears would fall from my face in Tree's garden. The stony corals were dying. During my time there, every day was another that SCTLD triumphed. The dive team was discouraged. Even when doing the hopeful work of restoration, planting new, branching corals grown in our nursery onto the reef, we felt disheartened. There were no visible or measurable signs that the work we did was making a difference. I felt disillusioned in my dream job. The days were long and I was bone-tired, but sleep can't mend a breaking heart. I needed a friend. And without words, Tree knew. I'm not sure he could see my tears, and we hardly spoke of the reefs, but when I sat in the garden, he would pass me a chilled glass of mint julep, made with leaves from the plant over his shoulder, and tell me another story.

One conversation about rare plants led me to mention phantom orchids. Tree had never heard of them, so I told the story of how several years back, I was hiking with a professor and group of students through the rich, mossy forests of Blakely Island. The foliage was thick overhead, and I trod through more shadow than sun, never seeing the pink blossoms for which I searched among the dark soil patched with moss. I was on the hunt for *Calypso bulbosa*, fairy slipper, a tiny and vibrantly pink species of wild orchid, with a speckled face and yellow throat.

Instead, like a ghost in the shadows, there sprang up a single pale bloom woven into the emerging roots of a spreading tree. The phantasmic stocks stood about a foot tall, the petals having tender tips and a pale-yellow center. The little plant was ethereal, as if a spotlight of mottled sun shone from above, highlighting it amidst all the darkness below.

My professor gasped and sprinted to the little cluster of blossoms. His eyes were wide as he thumbed through a plant taxonomy book, double-checking his own expertise before quietly declaring, "This is the extremely rare phantom orchid, *Cephalanthera austiniae*. It lives in symbiosis with a very specific family of fungus, which has a special association with only a few species of trees. In all my years of hiking this area, I have never seen one before." The group of students crowded around, careful to maintain a safe distance from the little novelty and its intricate root system.

Phantom orchids lie dormant underground for years at a time before deciding, when environmental conditions are ideal, to finally bloom.[1] They wait for the perfect time, when they and the world around them are ready. It was no wonder my professor had never seen one before. The orchid we saw may have waited up to seventeen years to bloom.[2] I admired the amazing flower, the fruit of years of laborious dormancy and growth, realizing that it was likely both the first and last phantom orchid I would ever see.

I thought the orchid's rarity, its strange ghostly uniqueness, so very similar in concept to a nursery coral. An underwater version of a tree nursery, a coral nursery borrows much of the same terminology. Instead of rows of baby plants in soil, our nursery had rows of coral "trees" anchored in sand on one end, buoyed to float on the other, with each tree holding up to one hundred baby *Acropora*, endangered branching corals. The trees were large constructions of crisscrossed PVC pipe lined with monofilament threads from which corals hung, much like ornaments on a Christmas tree. Starting as inch-long fragments broken off wild parent colonies, the baby corals were hung to grow through the seasons.

From their nursery perch, baby corals dangled in the midst of territorial hogfish and meandering conchs. They weathered the powerful torrents and frothing currents of tropical storms. They quadrupled in size and were eventually snipped from the trees with pliers by divers like myself. Placed in easily transportable milk crates, loads of corals were transported to various historical sites on the reef, locations where branching corals existed fifty years ago but had since died off. We secured these corals to the substrate, experimenting with different outplanting methods (ways of planting the coral), such as using nails and zip ties or two-part epoxy. Hauling heavy supplies and wielding hammers made for hard underwater labor. Once we'd firmly secured the corals, we wished them luck in their new home on the reef. In seasons to come, we would check up on the outplanted corals, measuring their survival and growth, always hoping to someday see the reef restored, looking like it had in years long passed. But change takes time; bigger changes sometimes take longer than we can or care to wait.

Back in the nursery, the place of stillness where growth occurred, the rows of coral trees through the hazy water looked much like a freeze-frame of an alien army in a sci-fi show. The thousands of baby corals, clinking gently together like wind chimes in the current and growing so quickly that the tips of their branches could fuse together, were a testament of our restoration efforts. It illustrated the lengths to which humans will go with their time and industry and creativity to fix what has been lost and broken. It was a tangible reality of the hope we held for the future.

Nowadays coral restoration is increasingly controversial. Experts and novices alike debate whether coral nurseries and outplanting efforts are viable, worthwhile, and scalable solutions to the problems at hand—reefs actively dying or already destroyed due to disease, changing water temperatures, human overuse and abuse, and other stressors.[3] Restoration efforts certainly have a time and place, but disagreement lies in whether the proposed solutions are the end-all be-all or a mere Band-Aid.

Some people treat restoration like the only solution that matters. This is hard to swallow for those who have seen long-term outplant monitoring result in high mortality, or for the people who recognize that restoration works only if the initial problems are addressed. For them, restoration can feel like an imprudent though well-intentioned waste. Courage and hard work alone will not solve our greatest problems if rationality and strategic, long-term planning are omitted. If corals are struggling because their environment is deteriorating, maybe we should focus on improving that environment before adding in more corals. Otherwise, we may be sentencing them to death.

Inversely, some people treat restoration like a complete waste. This is impossible to accept for those who truly understand what corals contribute environmentally and economically, or for people who have crunched the numbers and believe the risk of investment is far and away worth the cost. It is also hard to dismiss restoration when it is the simplest, most doable solution at hand, becoming steadily easier with advancing techniques and a growing pool of knowledge. In the face of rapidly deteriorating wild reefs, some experts such as Dr. David E. Vaughan, the founder and president of Plant A Million Corals Foundation, have staked the future success of reefs on the practice of active restoration.[4] We may not be able to create the perfect conditions needed for ideal coral growth, but we can at least give corals a shot! Simply put, we can plant as many of them as possible, as carefully as we can, and see what happens.

I argue that the restoration efforts with the greatest likelihood of long-term success are those built on a diverse foundation of research and researchers. But even among restoration advocates, there can be dissension. The way we talk about which species of corals should be prioritized, when and where to plant them, specific methods for growing and outplanting, who is doing the most (or best) work, and how much it matters . . . these can seem like moral topics. It is so easy to lose sight of the main issue when every molehill is somebody's mountain. As with everyone, scientists are flawed people. We are doing our best to understand and fix the flawed world. It shouldn't be surprising if some of our tools and methods are eventually found to be flawed too.

It reminds me of the advice given by Walt Jaap, the marine biologist who pioneered coral science at the national park where I worked, whose old publications and long-distance suggestions I came to rely on.[5] We never met in person. Corresponding through email, Walt told me, "Be open, but use your science to filter the scene. Be aware that work may be frustrating because something always happens—weather, gear problems, other things. Be willing to make changes. . . . We are all in this together, and I am pulling for you."[6] He gave me his phone number.

Besides our mutual love for corals and the old Canadian sitcom *The Red Green Show*, it turns out that Walt and I also had a friend in common: Tree. While we were sitting on the back patio one evening, I asked him to describe the Walt he knew.

Walt, Tree told me, was a "stereotypical spectacled scientist, always explaining the details of stuff I didn't really understand. He enjoyed tending corals, like you." He smiled. "I've always been more partial to plants." After that, Tree told me stories of a younger Walt. I imagined what it would have been like to be on a boat with those guys. Although restoration partnerships wouldn't come on the scene for another thirty years, I'm sure people argued over their work back then too.

The hardest parts of earthly restoration work are the same as that of spiritual restoration. We labor long and the process is slow. New challenges arise, and growth takes time. It can feel heartbreaking in the moment, and we don't necessarily learn the reasons for why bad things happen. We might die before seeing the outcome. We might wonder whether there is any value in our effort. Sometimes it's just work. But as A. W. Tozer encouraged, "Let every man abide in the calling wherein he is called and his work will be as sacred as the work of the ministry. It is not what a man does that determines whether his work is sacred or secular, it is *why* he does it. . . . Let a man sanctify the Lord God in his heart and he can thereafter do no common act."[7]

The work we do can be made sacred when done from a heart of desiring to please and pursue God, whether that is active restoration work or something else entirely. Today, the trees our great-grandfathers planted are ten times the size they knew, and we praise the Creator from within their shade. In one hundred years, perhaps a coral I planted will be the size of a Volkswagen Beetle and comparable to a great-grandmother. There is hope in that. This is why people pursue restoration. The first challenge is that the work is hard; the second challenge is accepting that perhaps we will never know the outcome of our efforts.

But we do not labor in vain. Not in our work as biologists, not in our lives as Christians. The life, death, and resurrection of Jesus make a different perspective possible. Because of this, "we do not lose heart. Though outwardly we are wasting away, yet inwardly we are being renewed day by day. For our light and momentary troubles are achieving for us an eternal glory that far outweighs them all. So we fix our eyes not on what is seen, but on what is unseen, since what is seen is temporary, but what is unseen is eternal" (2 Corinthians 4:16-18, NIV). We focus our attention not on the suffering of life or the coming of death, but on the infinite, restored life that awaits us. We realize that even as outwardly coral reefs around the globe are dying, yet inwardly God is in the business of redeeming and restoring day by day.

Maybe like corals, dangling on monofilament in the turbid waters of a nursery, we undergo substantial growth and preparation for our future lives without knowing it. For how is a baby *Acropora* to know, while it flails at the mercy of a current, that it is being nurtured in a nursery for a life of adventure in the wild garden? Like a phantom orchid lying dormant beneath soil for some seventeen years, perhaps the most beautiful developments in our souls take time. For how can a plant that has spent its whole life underground fathom the splendorous light that will someday grace its gorgeous new petals? Sometimes we don't see the outcome of our labor, but that doesn't mean God isn't already using it—for his purpose, for his glory, for his incredible restoration work that will someday be revealed within and to us in the promised Garden City.

I told Tree in the garden. When our friend, Walt Jaap, passed away, he did not know the profound influence he had on my perception of coral work. I was standing on the calcium carbonate foundation his career had laid, but I hadn't ever told him what that meant to me. Tree and I shared a quiet moment, remembering our friend, before he suddenly stood.

"You see that cactus behind you? It started off about four inches tall, and now look at it. It's touching the roof."

I turned around in my seat and chuckled at the prickly monstrosity. Of course, Tree would own a tall, lanky, prickly plant like that. "What kind of cactus is it?" I asked.

"Peruvian apple cactus. They only bloom at night when nobody's watching. One night per bloom. Gone in the morning."

"Looks like yours has some buds!"

"That's why I'm showing you. I think it's set to go off tonight. If you want to come back by around midnight, it'll be flowering."

I crept out of the team house around midnight and tiptoed across the lawn, strewn with dry leaves, to the hole in the wall of Tree's garden. Stealthily, I shone a light on the tall cactus and smiled. Enormous, cantaloupe-sized white flowers stared back at me. That such a strange, unfriendly plant could produce these incredible blooms seemed like a metaphor I couldn't quite grasp at the time. But a month later, I understood.

I had just come back to the island after a long, tiring day of outplanting corals from the nursery to the reef. Dragging my feet toward the house, I saw Tree in the distance by the front porch, scuffing at the dirt with one foot and holding something tall and green in hand. I noticed his arm had a long, bloody scratch.

"What are you doing, Tree?"

He glanced at me from the corner of his eye and put his free hand on his hip. After a long moment, he gestured at what he had just planted. "For you." Immediately recognizable, it was a large Peruvian apple cactus propagule about five feet tall with no less than six buds, cut from his own garden. It would bloom that night.

Tears sprang into my eyes at the realization of the gift, a testament of our friendship. Prickly, potentially painful, but beautiful and worth the wait. Something I would see every day when I came home from work. When I asked how he'd managed the surprise, Tree answered reluctantly, "It was tricky. And I wasn't sure of the timing for when you'd be back from diving. Took a few weeks to plan and a couple of hours to get cut down."

"Tree, I think this is the nicest thing anyone has ever done for me. Thank you."

The cactus bloomed many times in the months after that, always at night, only to be seen by my eyes. I drank in the sight of the enormous white blooms with hope, knowing they would not survive to morning, meditating on what is eternal. "The grass withers and the flowers fall, but the word of our God endures forever" (Isaiah 40:8, NIV). Staring at the marvelous beauty of all that cactus's hard growth and my friend's careful labor, I thought of my own work. I crawled into my sleeping bag those nights and dreamed of flowers being ghosts and corals as wind chimes. I awoke in the morning and worked beside my team, laboring fully in the work of the Lord, knowing that regardless of how tired my body was after diving or whether the outplants survived or SCTLD was eradicated, none of it was in vain.[8]

As Walt had advised, I stayed open to change. When new research suggested we change our outplanting strategies, we did. While Tree cared for his plants, finding the best spots of sun and shade for them to grow, I planted corals, searching out the places where they would hopefully thrive for generations to come, although I might never live to see them. Regardless of whether or not this will mean something to somebody yet to be born, God will use my efforts. I must simply do what he has called me to do, dark and earthly and prickly as it may be, tending these wild gardens.

EPILOGUE

the greatest hope

It has been over a year since I felt a sea breeze tangle my hair. Like uprooted kelp, my toes have craved sand, and my eyes have longed for the sight of that rolling blue wilderness. Joyfully bursting from the forest trail and clambering over the array of driftwood, I run to my old friend—the ocean. My husband is close on my heels, laughing at my joy. Since he met me, the ocean has become his friend too. He is particularly delighted by tide pools, calling me over to observe a cluster of bejeweled sea stars, to catch an especially feisty crab, or to name a peculiar creature he is encountering for the first time. Like the tide and the creatures it covers, life teems with change. But the feelings I have and the questions I ask at the ocean are the same.

During my time at Dry Tortugas National Park, the Florida Reef changed. SCTLD's ravages left half-living fragments and skeletons of hard corals

where once they had thrived. Initially, we wondered if the disease would burn itself out, like a wildfire. In reality, the disease became endemic, meaning that it became so common that it is now considered a permanent fixture of that ecosystem. For reasons I chalk up to stubborn denial and heartrending nostalgia, scientists continue to classify the Florida Reef as a barrier reef—wall-like in structure. SCTLD broke down that wall into staggered, pockmarked pieces. If we did not know its historical form, the Florida Reef would be considered a series of patch reefs—tiny, struggling, and yet tremendously important.

Academic, research, and government agencies are still trying to determine what caused the disease and whether the antibiotic treatment worked. As with all human endeavors, sometimes science fails. In my personal opinion, the fact that we continue to ask, "Did it work?" suggests that it did not. Long-term monitoring data yet to be collected will tell the full tale, but the current lack of surviving reef is certainly not in favor of the antibiotic treatment.

While this may be the end of a story, it is far from the end of *the* story. Around the world, there are dedicated, hardworking people endeavoring for the good of coral reefs and the communities they support. I have had the privilege of shaking hands with some and working alongside others. We may have differing backgrounds, perspectives, and opinions, but one thing remains certain: We wouldn't be doing what we do if there wasn't at least some small glimmering fragment of hope.

There are some biological possibilities—hope that can be found through scientific determination and discovery. On the Florida Reef and beyond, disease intervention efforts have transitioned to restoration practices. Coral biologists are raising corals like agricultural crops in on-land facilities and underwater nurseries and outplanting them to reefs as quickly as possible.[1] Researchers are breeding and testing different kinds of corals in an attempt to propagate genetic strains that are better equipped to handle the extreme changes we predict will continue to happen in our oceans.[2] Aquarists are gathering large-scale collections of coral fragments in a Noah's ark sort of library, hoping to preserve as much biodiversity as possible until scientific innovation can catch up with the changing environment and present a reliable long-term solution.[3] Today, there are pieces of different kinds of corals (including those affected by SCTLD) preserved in various aquariums, museums, and research facilities throughout the world. During my time at Dry Tortugas, fragments of the endangered *Dendrogyra cylindrus* castles were collected in a "rescue mission."[4] Last I heard, these fragments were alive, flourishing, and cherished in Miami's Frost Museum of Science, which they now call home. And though it is small beans compared to these efforts, my quiet hope is that by the end of my lifetime, some of the corals outplanted in the stories of this book will be thriving and growing as their ancestors once did.

For those of us who are Christians, we ought to have some faith in science and even greater faith in the saving power of Jesus Christ. He has given a bold, shining promise regarding the restoration that is to come, and thus we possess the greatest hope even while knowing that we—both corals and people—are living in broken places within a suffering world, waiting for the eternal restoration that has been promised to us.

Even in the darkest maze on a diseased reef, we can experience the presence of the God who loves and provides for and offers promises to every created thing. He made Porky the porcupinefish beautifully and wonderfully. He provided for the great-grandmother coral, giving her wisdom, longevity, and a peaceful end. He loves the Magic Castles far better than I ever could. It is through God that we all move and breathe and exist.[5]

Since my final days at Dry Tortugas, I am occasionally asked by kindhearted individuals, "What can I do to help corals?" While there are a slew of online resources providing excellent suggestions for the day-to-day ways you can help corals (recycle, conserve water, carpool, avoid pesticides, support coastal communities, fund research and restoration, etc.), it may be unsurprising that my answer leans more philosophical. The best thing you can do to help corals is to love them.

I am convinced that people protect what they love. But we can only learn to love what we understand. By taking time to learn about corals, you are already making leaps and bounds toward loving them. And by loving them, it may be possible to save them, even within our lifetimes.

If God so loved the world—corals included—then we must as well! With all of creation, we rejoice, weep, and labor. We wrestle with questions to find the Truth, to really see God. And when the waters run deep and dark, we remember that he is actively working, even through us, to redeem and restore all things.

The corals taught me many things. Among them, that joy and suffering can exist side by side. Birthdays and death days are bookends to our stories. Endings and beginnings go together. Just so, while the reefs were dying, a part of my life grew that had not existed before.

I had always planned to go where the corals were, but after a long season of chasing them, God surprised me by leading in a different direction—inland. The corals were leaving the world just as an amazing person came into mine. Together, he and I decided to dive into a deep, unexplored thing called marriage.

That is also how this book came to be. Becoming a landlocked marine biologist with a tremendously supportive life partner created the ideal niche for these thoughts turned words to flourish and grow. Turns out, God needed to take me away from the ocean in order for me to write about it.

I don't know what will happen next or when I'll get to race sea-foam down the beach again. But today I am satisfied in the presence of God, catching glimpses of him between tide pools, running out with the surf, glorying in the praise of the crashing waves. Creation lifts its multitude of voices to honor our Creator. The fish rejoice, Leviathans frolic, and the corals sing. I am overcome by the work of God's hands and his great desire for us, chasing over the expanse, closing the gap, and meeting us in the depths.

acknowledgments

Sometimes it seems as though my whole life has been spent collecting evidence that God is able "to accomplish infinitely more than we might ask or think" (Ephesians 3:20). This book is one example of that. The people God has intertwined with my life are another.

A ton of credit goes to my parents, who knew long before I did that I'd write a book someday. They introduced me to two great loves—Jesus and the ocean—and taught me how to navigate questions, fear, doubt, and wonder. My brother, Sam, who is a fantastic listener and adventure buddy, helped me learn the art of storytelling. To use his words, we go together like eggs and peanut butter. This book would not have happened in this time, in this way, were it not for my husband, Clint. He is my best friend and greatest blessing in life.

Many dear friends have been sounding boards for the development of this book. To Alicia Manfroy, Coral Reid, Lindsey Lemke, Marissa Myer, Max Kam, Mickey Chan, Tree Gotshall, and Will Demarest—you were the friends walking beside me in these stories, and yours were the faces I envisioned sitting across from me as I wrote.

I owe a massive thank you (and hug) to Sherri Wilson Johnson, my agent, who is an absolute joy to work with and who believed wholeheartedly in this book from the start. Rachelle Gardner of Gardner Literary has been a fountain of wisdom on the path to publication.

I want to thank Kara Leonino, who, from the get-go, understood this book's message and has championed it. Jan Long Harris, Porky's number-one fan, opened many additional doors of opportunity for this book. Powerhouse editor Kim Miller helped my writing live up to its full potential, graciously enduring my nerdy tangents and questions. Elizabeth Czajkowski worked hard to ensure that everything reads smoothly and correctly. Thank you to acquisitions editors Mary Campbell and Kaylee Small for helping bring this book to publication. I also want to thank Michelle Polsley and Brianna Coyle for their tireless work in marketing and production. Dean Renninger and Julie Chen delivered the vibrant, dynamic visuals that every book about marine science deserves. I am so very grateful for everyone at Tyndale House who helped make this book a reality!

Having lived in lots of places, I have been a part of many different churches, but two stand out: Townsville Wesleyan Methodist Church in Queensland, Australia, and LifePoint Church in Nebraska, USA. I will never forget Pastor Stuart Hall fixing my motor scooter and praying for my future. The Leong family, in particular, ministered to me on their succulent-covered veranda with generosity, kindness, and many cups of tea. Thank you to the TWMC young adults group, who helped me explore theological questions and create lasting memories. The congregation of LifePoint has offered unwavering support and encouragement for this book. For a triple-landlocked state, they sing a lot of worship songs that reference the ocean, for which I am grateful. I am especially thankful to Pastor Chris Winegar for his notes on "being charitable" and for theological checks to my early manuscript.

My teachers helped form me as both a scientist and Christian. I will never stop thanking Eric Long, who has answered my endless questions with empathy and humor; I aspire to be more like him. Jennifer Tenlen has been both a mentor and friend, providing some of the sweetest and soundest advice I have ever received. I'm still not over diving the Blue Hole with Timothy Nelson, the coolest phycologist ever. I am always elated to catch up with Mike McCormack, who so kindly continues to cheerlead and celebrate my career trajectory. Heather Solsvik was one of the first to tell me I was good at writing, and she taught me to be better. I also want to thank David Bourne for the incredible research gigs, career advice, and cappuccinos. And lastly, thanks to Ryan Ferrer for helping teach me to trust God's will for my life.

discussion questions

1. What is your favorite story in the book? Which characters stand out to you most, and how have they impacted your view of creation?
2. What spiritual or scientific questions are you afraid to ask? What is the risk of not asking them?
3. Provide some examples of the ways in which faith and science interact. How might you approach areas where they initially seem incompatible?
4. There are four different views of how creationism functions: young earth, old earth, progressive, and evolutionary. Which of the four do you find most compelling and why?
5. What did you learn from this book about marine life? What aspects of the underwater world amaze you most? Would you prefer to go tide pooling or scuba diving? Why?

6. What are some examples of current fears or challenges that you might describe as “Leviathans”? How does remembering God’s sovereignty shift your perspective of these things?

7. God gave humankind the mission to “serve and protect” creation. How does this impact your understanding of the world? How does it apply to your life?

8. Describe the connection between creation and Creator. How does studying creation help you understand God better?

9. Stony coral tissue loss disease (SCTLD) dramatically shifted the Florida Reef’s ecosystem. Can you think of a comparable event that changed another ecosystem? What was the impact of that event?

10. Underwater labor is difficult. What are some aspects of your life that feel laborious? When you consider what God has called you to do and that he equips you, how does your approach to these responsibilities change?

11. Suffering and death are outcomes of sin in the world, affecting all of creation. How do you view suffering and death? How does thinking of Jesus transform your outlook?

12. How do you imagine the Kingdom of Heaven? What kinds of activity and ecology are present?

13. Who might you describe as a “dive buddy” in your life? What characteristics do you most appreciate about them?

14. Describe your relationship with God. What are some practical steps you could take to deepen that relationship, growing in greater intimacy by drawing near like Mister Snuggles, the seahorse from chapter 13?

15. Have you ever experienced a dream being delayed or derailed? Have you ever felt like God was hidden from sight in the "Maze"? If so, what did you learn from these experiences?

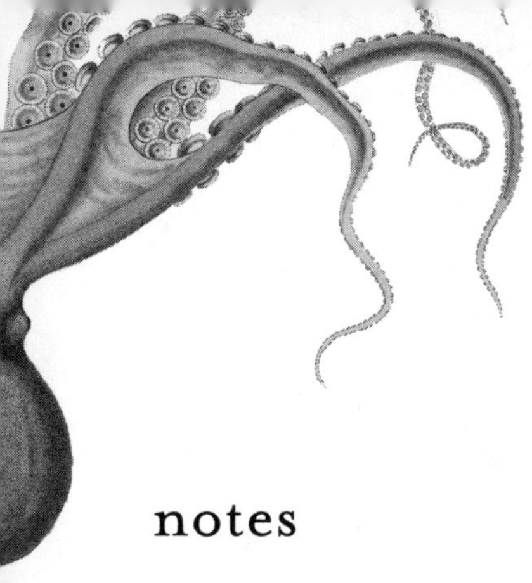

notes

PREFACE: UNFATHOMABLE SEAS

1. Francis, encyclical letter *Laudato Si'*, May 24, 2015, https://www.vatican.va/content/dam/francesco/pdf/encyclicals/documents/papa-francesco_20150524_enciclica-laudato-si_en.pdf.

CHAPTER 1: LAND OF THE LIVING

1. "The Windjammer Wreck," National Park Service, last updated April 11, 2016, https://www.nps.gov/drto/learn/historyculture/windjammer.htm.
2. "Stony Coral Tissue Loss Disease Found at Dry Tortugas National Park," National Park Service, June 2, 2021, https://floridascoralreef.org/2021/06/02/stony-coral-tissue-loss-disease-found-at-dry-tortugas-national-park; "Investigating Coral Disease Transmission and Mitigation Strategies," Atlantic Oceanographic and Meteorological Laboratory, accessed June 6, 2024, https://www.aoml.noaa.gov/coral-disease.
3. "Stony Coral Tissue Loss Disease (SCTLD)," Coral Disease and Health Consortium, accessed September 25, 2024, https://cdhc.noaa.gov/coral-disease/characterized-diseases/stony-coral-tissue-loss-disease-sctld; Lorenzo Alvarez-Filip et al., "Stony Coral Tissue Loss Disease Decimated Caribbean Coral Populations and Reshaped Reef Functionality," *Communications Biology* 5, 440 (2022), https://doi.org/10.1038/s42003-022-03398-6; "Stony Coral Tissue Loss Disease (SCTLDS)," The Caribbean Oceanic Restoration and Education Foundation, accessed September 20, 2024, https://www.corevi.org/stony-coral-tissue-loss-disease.html.
4. Tiffany Duong, "Stony Coral Tissue Loss Disease Reaches Dry Tortugas; Park Service Responds," *Keys Weekly*, June 14, 2021, https://keysweekly

.com/42/stony-coral-tissue-loss-disease-reaches-dry-tortugas-park-service-responds.

5. Erin Papke et al., "Stony Coral Tissue Loss Disease: A Review of Emergence, Impacts, Etiology, Diagnostics, and Intervention," *Frontiers in Marine Science* 10 (January 25, 2024), https://doi.org/10.3389/fmars.2023.1321271.
6. Milton Muldrow Jr., Edward C. M. Parsons, and Robert Jonas, "Shifting Baseline Syndrome among Coral Reef Scientists," *Humanities and Social Sciences Communications* 7, no. 42 (July 2020), https://doi.org/10.1057/s41599-020-0526-0. See also Lauren T. Toth et al., "The Past, Present, and Future of Coral Reef Growth in the Florida Keys," *Global Change Biology* 28, no. 17 (July 2022): 5294–5309, https://doi.org/10.1111/gcb.16295.
7. "Basic Information about Coral Reefs," United States Environmental Protection Agency, last updated February 28, 2024, https://www.epa.gov/coral-reefs/basic-information-about-coral-reefs. See also "Why We Need Coral Reefs," Great Barrier Reef Foundation, April 22, 2024, https://www.barrierreef.org/news/news/why-we-need-coral-reefs.
8. Hollister et al., "Stony Coral Tissue Loss Disease (SCLTD) Reconnaissance, Intervention, and Monitoring in Dry Tortugas National Park," (June 15, 2022), Florida Department of Environmental Protection, https://floridadep.gov/sites/default/files/10.%20Stony%20Coral%20Tissue%20Loss%20Disease%20%28SCTLD%29%20Reconnaissance%2C%20Intervention%2C%20and%20Monitoring%20in%20Dry%20Tortugas%20National%20Park_0.pdf.
9. Duong, "Stony Coral Tissue Loss Disease Reaches Dry Tortugas."
10. C. S. Lewis, *The Four Loves* (New York: Harcourt Brace Jovanovich, 1960), 169.
11. John Muir, *Stickeen: The Story of a Dog* (Boston: Houghton Mifflin, 1909), 17.
12. Timothy Keller, "Christian Hope and Suffering" from *Living in Hope* sermon series, May 16, 2024, in *Gospel in Life*, podcast, MP3 audio, 28:00, https://gospelinlife.com/sermon/christian-hope-and-suffering.
13. Keller, "Christian Hope and Suffering," 29:16.
14. Keller, "Christian Hope and Suffering," 31:07.
15. Revelation 21:5.
16. Romans 5:17; 1 Corinthians 15:21-22.
17. Rainer Maria Rilke, *Rilke's Book of Hours: Love Poems to God* (New York: Riverhead Books, 2005), 119.
18. This beautiful name for the home God is preparing for his people has been discussed by many Christian leaders, including Timothy Keller, "The Garden—City of God," sermon, Redeemer Presbyterian Church,

Manhattan, New York, April 26, 2009; and John Mark Comer, *Garden City: Work, Rest, and the Art of Being Human* (Grand Rapids, MI: Zondervan, 2015).

CHAPTER 2: THE BOTHERS OF BEACHCOMBING

1. Edward O. Wilson, *Genesis: The Deep Origin of Societies* (New York: Liveright Publishing, 2019), 32–33.
2. Yann Martell, *Life of Pi* (Orlando, FL: Harcourt, 2001), 30.
3. Bruce L. Shelley, *Church History in Plain Language* (Waco, TX: Word Books, 1982), 116.
4. Phylicia Masonheimer, *Every Woman a Theologian: Know What You Believe. Live It Confidently. Communicate It Graciously.* (Nashville: Thomas Nelson, 2023), 52.
5. "How Is Evolutionary Creation Different from Evolutionism, Intelligent Design, and Creationism?," BioLogos, last updated November 20, 2022, https://biologos.org/common-questions/how-is-biologos-different-from-evolutionism-intelligent-design-and-creationism.
6. Ted Davis, "Old-Earth (Progressive) Creationism: History and Beliefs," BioLogos, June 19, 2012, https://biologos.org/series/science-and-the-bible/articles/old-earth-progressive-creationism-history-and-beliefs.
7. John C. Lennox, *Seven Days That Divide the World: The Beginning According to Genesis and Science* (Grand Rapids, MI: Zondervan, 2011).
8. John H. Walton, "Proposition 14: God's Roles as Creator and Sustainer Are Less Different Than We Have Thought," *The Lost World of Genesis One: Ancient Cosmology and the Origins Debate* (Downers Grove, IL: IVP Academic, 2009); "How Is Evolutionary Creation Different?," BioLogos.
9. For more, see How to Read the Bible Series: Literary Styles, The Bible Project, June 22, 2017, https://bibleproject.com/explore/video/literary-styles-bible.
10. "How is Evolutionary Creation Different?," BioLogos.
11. J. Richard Middleton, "How Should We Interpret Biblical Genealogies?," BioLogos, July 28, 2021, https://biologos.org/series/how-should-we-interpret-biblical-genealogies; David A. Falk, "Understanding Genealogies in the Bible (Part 1)," Egypt and the Bible, February 19, 2019, https://www.egyptandthebible.com/index.php/2019/02/19/understanding-genealogies-in-the-bible-part-1/.
12. "The Genealogical Adam and Eve," episode 7 of the *Ancient Cosmology* series in the *BibleProject* podcast, July 5, 2021, produced by Cooper Peltz, Dan Gummel, and Zach McKinley, https://bibleproject.com/podcast/genealogical-adam-and-eve.
13. For more, see Aaron Demsky, "Reading Biblical Genealogies," TheTorah.com, 2016, https://www.thetorah.com/article/reading-biblical-genealogies.

14. Tremper Longman III, *Genesis*, The Story of God Commentary (Grand Rapids, MI: Zondervan, 2016), 94.
15. Kenneth A. Mathews, "Are the Biblical Genealogies Reliable?," *The Apologetics Study Bible* (CSB) (Nashville: Holman Bible Publishers, 2018), 13, https://www.accordancebible.com/apologetics-study-bible-profile.
16. Mathews, "Are the Biblical Genealogies Reliable?"; Demsky, "Reading Biblical Genealogies."
17. W. Creighton Marlowe, "Patterns, Parallels, and Poetics in Genesis 1," *The Journal of Inductive Biblical Studies* 3, no. 1 (Winter 2016): 6–27, https://place.asburyseminary.edu/cgi/viewcontent.cgi?article=1053&context=jibs. See also J. Richard Middleton, "What is the Relationship Between the Creation Accounts in Genesis 1 and 2?," BioLogos, January 3, 2018, https://biologos.org/articles/what-is-the-relationship-between-the-creation-accounts-in-genesis-1-and-2.
18. Terry Mortenson and Thane H. Ury, eds. *Coming to Grips with Genesis: Biblical Authority and the Age of the Earth* (Green Forest, AR: New Leaf Publishing Group, 2008), 77.
19. Walton, *The Lost World of Genesis One*.
20. Francis, encyclical letter *Laudato Si'*, May 24, 2015, https://www.vatican.va/content/dam/francesco/pdf/encyclicals/documents/papa-francesco_20150524_enciclica-laudato-si_en.pdf.
21. Charles Colson, *God and Government: An Insider's View on the Boundaries between Faith and Politics* (Grand Rapids, MI: Zondervan, 2007), 78.

CHAPTER 3: CHOIR OF CREATION

1. Quoted in foreword to David E. Vaughn, ed., *Active Coral Restoration: Techniques for a Changing Planet* (Plantation, FL: J. Ross Publishing, 2021), xv.
2. A. W. Tozer, *The Pursuit of God* (Floyd, VA: Rediscovered Books, 2014), chapter 5.
3. Catechism of the Catholic Church, section 2, paragraph 5, "Heaven and Earth"; quoted by Pope Francis, encyclical letter *Laudato Si'*, May 24, 2015, https://www.vatican.va/content/dam/francesco/pdf/encyclicals/documents/papa-francesco_20150524_enciclica-laudato-si_en.pdf.
4. George MacDonald, *Phantastes* (Grand Rapids, MI: William B. Eerdmans Publishing Co., 2000), 77.
5. Coring trees is a way to determine how old a tree is without damaging it. A straw-like mechanism is screwed into the living tree and then pulled out. It extracts a thin layer of the tree, from its outermost (youngest) rings to its innermost (oldest) rings. The number of rings corresponds to the tree's age. The sample also offers valuable insights into the history of that tree's life.

Tree cores testify to seasons of drought, extra-long winters, and occasional climate-related events.

6. John H. Walton, *The Lost World of Genesis One: Ancient Cosmology and the Origins Debate* (Downers Grove, IL: IVP Academic, 2009), 83–84.
7. Walton, *The Lost World of Genesis One*, 83.
8. Genesis 1:4, 9-10, 12, 18, 21, 25.
9. C. S. Lewis, *Present Concerns* (San Diego: Harvest Books, 2002), 79.
10. C. S. Lewis, *The Four Loves* (New York: HarperOne, 2017), 25–26.
11. Revelation 21:1.
12. See, for example, Leviticus 10:1-3 and 1 Samuel 4.
13. 2 Chronicles 20:21; 1 Chronicles 16:34; Psalm 118:1; Psalm 136.
14. Francis, *Laudato Si'*.

CHAPTER 4: LEVIATHANS IN THE DEEP

1. Li Cohen, "'Incredibly Rare' 180-Million-Year-Old Giant 'Sea Dragon' Fossil Discovered in U.K.," CBS News, last updated January 20, 2022, https://www.cbsnews.com/news/sea-dragon-ichthyosaur-fossil-rutland-uk.
2. "What Was the Leviathan?," Got Questions, accessed June 13, 2024, https://www.gotquestions.org/leviathan.html.
3. For Leviathan as sea creature, see Job 3:8; 41:1; Psalm 74:14; 104:26. For Leviathan as oppressive earthly power or Satan, see Isaiah 27:1; Revelation 13:1-2.
4. Julie Dillon, *Beneath the Surface*, The Art of Julie Dillon, 2021, https://www.juliedillonart.com/beneath-the-surface.
5. Though the word *Leviathan* isn't used in Revelation, some commentators point out the similarities between the chaos caused by the giant sea creature in the Old Testament and the dragon in Revelation. See, for example, Hope Bolinger, "What Is the Leviathan according to the Bible?," Christianity.com, updated June 12, 2024, https://www.christianity.com/wiki/bible/what-is-the-leviathan-according-to-the-bible.html.
6. Job 3:8; 41:1; Psalm 74:14; 104:26; Isaiah 27:1; 51:9.
7. "Great Belize Blue Hole," Belize.com, last updated May 15, 2024, https://belize.com/belize-blue-hole.
8. Khaila Gentle, "Dive Magazine Celebrates 80 Years of Jacques Cousteau's Favourite Dive Sites—The Great Blue Hole Included," *Caribbean Culture and Lifestyle*, accessed June 13, 2024, https://caribbeanlifestyle.com/dive-magazine-80-years-jacques-cousteaus-favourite-dive-sites.
9. Because underwater terrain can be so variable, depth is always measured as distance from the surface.
10. Robert Kurson, *Shadow Divers: The True Adventure of Two Americans Who Risked Everything to Solve One of the Lost Mysteries of World War II* (New York: Random House, 2005), 22–24.

11. Kurson, *Shadow Divers*, 38.
12. Genesis 1:20, 31.

CHAPTER 5: LIVING WATER

1. Dr. Michael A. Milton, "When Was the Bible Written? The History of God's Word," Bible Study Tools, last updated October 23, 2023, https://www.biblestudytools.com/bible-study/topical-studies/when-was-the-bible-written.html.
2. For more on this, see Matthew Barrett, "The Authority and Inerrancy of Scripture," The Gospel Coalition, accessed September 13, 2024, https://www.thegospelcoalition.org/essay/authority-inerrancy-scripture. See also Phylicia Masonheimer, *Every Woman a Theologian: Know What You Believe. Live It Confidently. Communicate It Graciously.* (Nashville: Thomas Nelson, 2023), 12–16.
3. For example, see Genesis 1:3.

CHAPTER 6: VALLEY OF THE SHADOW

1. As quoted by Jennifer Hackenbruch, "For Your Spirit: Honoring Your Journey," Hospice of North Idaho, 2021.
2. Switchfoot, "What It Costs," written by Tim Foreman, track 4 on *The Edge of the Earth*, Penny Farthing Music, 2014.
3. Neil deGrasse Tyson, *Astrophysics for People in a Hurry* (New York: W. W. Norton, 2017), 32–33.
4. Charles Darwin, *On the Origin of Species: By Means of Natural Selection or the Preservation of Favoured Races in the Struggle for Life* (New York: Penguin Classics, 2009), 427.
5. This is why we do not need to be offended by "survival of the fittest." Hypothetically, if this principle were philosophical and we were meant to apply natural selection to how we live, this would be ableist and evil. But natural selection is a scientific process, not a moral philosophy. It tells us *how* the world works, not how the world *should* work. And interestingly, natural selection is accompanied by some compelling instances of altruism in nature. For one example, there are many notable instances where humpback whales have intervened to rescue seals from orcas.
6. John H. Walton, *The Lost World of Adam and Eve: Genesis 2–3 and the Human Origins Debate* (Downers Grove, IL: IVP Academic, 2015), 13.
7. Aldo Leopold, "Thinking like a Mountain," in *A Sand County Almanac: And Sketches Here and There* (New York: Oxford University Press, 2020).
8. William J. Ripple and Robert L. Beschta, "Trophic Cascades in Yellowstone: The First 15 Years after Wolf Reintroduction," *Biological Conservation* 145, no. 1 (January 2012): 205–213, https://doi.org/10.1016/j.biocon.2011.11.005.

9. Jim Robbins, "Yellowstone's Wolves: A Debate over Their Role in the Park's Ecosystem," *New York Times*, April 23, 2024, https://www.nytimes.com/2024/04/23/science/yellowstone-wolves-elk-bison-climate-change.html.
10. John C. Bythell, Barbara E. Brown, and Thomas B.L. Kirkwood, "Do Reef Corals Age?," *Biological Reviews of the Cambridge Philosophical Society*, 93, no. 2 (May 2018): 1192–1202, Epub December 28, 2017, https://doi.org/10.1111/brv.12391.
11. Terence P. Hughes, David Ayre, and Joseph H. Connell, "The Evolutionary Ecology of Corals," *Trends in Ecology and Evolution* 7, no. 9 (September 1992): 292–295, https://doi.org/10.1016/0169-5347(92)90225-Z.
12. Adam Smith et al., "Field Measurements of a Massive *Porites* Coral at Goolboodi (Orpheus Island), Great Barrier Reef," *Scientific Reports* 11 (2021): 15334, https://doi.org/10.1038/s41598-021-94818-w.
13. There is some debate about whether the death that occurs due to sin is physical or spiritual. If physical, we are left pondering why the fossil record makes it clear that creation experienced death prior to the existence of humans. If the wages of sin is instead a spiritual death, this scientific conflict is resolved. This is a question I'm still wrestling with; for more background, visit Dave Miller, "Does the Bible Teach That Human Death Is the Result of Sin?," BioLogos, June 10, 2019, https://biologos.org/articles/does-the-bible-teach-that-human-death-is-the-result-of-sin.
14. Genesis 2:16-17; Romans 6:23.
15. "What the Bible Says about Sin as Missing the Mark" (from *Forerunner Commentary*), BibleTools.org, accessed July 1, 2024, https://www.bibletools.org/index.cfm/fuseaction/topical.show/RTD/cgg/ID/6773/Sin-as-Missing-Mark.htm.
16. Romans 5:12.
17. Genesis 2:17; see also Genesis 5:5.
18. Phylicia Masonheimer, *Every Woman a Theologian: Know What You Believe. Live It Confidently. Communicate It Graciously.* (Nashville: Thomas Nelson, 2023), 55.
19. Romans 3:23; 5:10.
20. Hebrews 12:2.
21. Acts 2:18.
22. Psalm 23:6; Romans 6:23.
23. Romans 8:14-16; 1 John 3:1.
24. 1 Corinthians 15:26.
25. C. S. Lewis, *A Grief Observed* (New York: Bantam Books, 1976), 72–73.
26. Ezekiel 37:1-14, NIV.

CHAPTER 7: CRAB AND CLAM TREASURES

1. 2 Corinthians 12:9.
2. Philippians 4:19.
3. Psalm 37:4.
4. 1 Peter 5:7.
5. Matthew 11:28-30.
6. Psalm 18:28, 32-34.
7. Hebrews 12:15.
8. Ephesians 4:31.
9. Pamela Soo and Peter A. Todd, "The Behaviour of Giant Clams (Bivalvia: Cardiidae: Tridacninae," *Marine Biology* 161, no. 12 (2014): 2699–2717, https://doi.org/10.1007/s00227-014-2545-0.
10. Soo and Todd, "The Behaviour of Giant Clams."
11. 1 Peter 5:6-7, NIV.
12. Romans 5:3-5, NIV.
13. Roberta Naas, "This $100 Million Pearl Is the Largest and Most Expensive in the World," *Forbes*, August 23, 2016, https://www.forbes.com/sites/robertanaas/2016/08/23/100-million-pearl-hidden-under-bed-sets-world-record-as-largest-most-expensive-pearl-in-the-world; Elle Hunt, "Fisherman Hands In Giant Pearl He Kept under the Bed for 10 years," *The Guardian*, August 23, 2016, https://www.theguardian.com/world/2016/aug/24/fisherman-hands-in-giant-pearl-he-tossed-under-the-bed-10-years-ago.

CHAPTER 8: INVISIBLE LIGHTS

1. Neil deGrasse Tyson, *Astrophysics for People in a Hurry* (New York: W. W. Norton, 2017), 85.
2. Hans E. Waldenmaier, Anderson G. Oliveira, and Cassius V. Stevani, "Thoughts on the Diversity of Convergent Evolution of Bioluminescence on Earth," *International Journal of Astrobiology* 11, no. 4 (October 2012): 335–343, https://doi.org/10.1017/S1473550412000146.
3. Matthew P. Davis, John S. Sparks, and W. Leo Smith, "Repeated and Widespread Evolution of Bioluminescence in Marine Fishes," *PLOS ONE* 11, no. 6 (June 8, 2016): e0155154, https://doi.org/10.1371/journal.pone.0155154.
4. "Factsheet: Bioluminescence," DeepOceanEducationProject.org, accessed September 23, 2024, https://oceanexplorer.noaa.gov/edu/materials/bioluminescence-fact-sheet.pdf.
5. Peter B. Moyle and Joseph J. Cech, *Fishes: An Introduction to Ichthyology* (Upper Saddle River, NJ: Pearson Prentice Hall, 2004), 585.
6. Astrid Kodric-Brown and Sally C. Johnson, "Ultraviolet Reflectance Patterns of Male Guppies Enhance Their Attractiveness to Females,"

Animal Behavior 63, no. 2 (February 2002): 391–396, https://doi.org/10.1006/anbe.2001.1917.

7. Tyson, *Astrophysics for People in a Hurry*, 187–188.
8. Gerald H. Jacobs, "Ultraviolet Vision in Vertebrates," *American Zoologist* 32, no. 4, (August 1992): 544–554, https://doi.org/10.1093/icb/32.4.544; George S. Losey, "Crypsis and Communication Functions of UV-Visible Coloration in Two Coral Reef Damselfish, *Dascyllus aruanus* and *D. reticulatus*," *Animal Behavior* 66, no. 2 (August 2003): 299–307, https://doi.org/10.1006/anbe.2003.2214; Yongsheng Shi and Shozo Yokoyama, "Molecular Analysis of the Evolutionary Significance of Ultraviolet Vision in Vertebrates," *Proceedings of the National Academy of Sciences* 100, no. 14 (July 2003): 8308–8313, https://doi.org/10.1073/pnas.1532535100.
9. Switchfoot, "Stars," written by Jon Foreman, track 2 on *Nothing Is Sound*, Sony BMG, 2005.
10. "About Hubble," National Aeronautics and Space Administration, accessed July 31, 2024, https://science.nasa.gov/mission/hubble/overview/about-hubble; Kieran P. O'Dea et al., "*Hubble Space Telescope* Far-Ultraviolet Observations of Brightest Cluster Galaxies: The Role of Star Formation in Cooling Flows and BCG Evolution," *The Astrophysical Journal*, August 2, 2010, https://doi.org/10.1088/0004-637X/719/2/1619.
11. "Webb's Orbit," National Aeronautics and Space Administration, accessed July 31, 2024, https://science.nasa.gov/mission/webb/orbit; P. Jakobsen et al., "The Near-Infrared Spectrograph (NIRSpec) on the *James Webb* Space Telescope: Overview of the Instrument and Its Capabilities," *Astronomy and Astrophysics* 661 (May 2022), https://doi.org/10.1051/0004-6361/202142663.

CHAPTER 9: BUDDY BREATHING

1. "The Trail: The 'Crown Jewel' of Rail-to-Trail Adventures," Route of the Hiawatha, accessed July 2, 2024, https://www.ridethehiawatha.com/the-trail.
2. To prevent serious injury, one of the top rules of scuba diving is that you never hold your breath. Doing so can allow gases to build up in the lungs, overexpanding and potentially bursting them. Even when a regulator is removed, it is important to continuously exhale.

CHAPTER 10: FIREWORMS AND BUTTERFLIES

1. Luna B. Leopold, ed., *Round River: From the Journals of Aldo Leopold*, (New York: Oxford University Press, 1953), 147.
2. Brian K. Walker et al., "Optimizing Stony Coral Tissue Loss Disease (SCTLD) Intervention Treatments on *Montastraea cavernosa* in an

Endemic Zone," *Frontiers in Marine Science* 8 (July 2021), https://doi.org/10.3389/fmars.2021.666224.
3. Genesis 3:17; Romans 6:23.
4. Isaiah 45:7.
5. Phylicia Masonheimer, *Every Woman a Theologian: Know What You Believe. Live It Confidently. Communicate It Graciously.* (Nashville: Thomas Nelson, 2023), 36–37.
6. C. S. Lewis, *The Problem of Pain* (New York: HarperOne, 2015), 36.
7. 1 Corinthians 13:4-8.
8. Lewis, *The Problem of Pain*, 39.
9. Ephesians 2:5.
10. Deuteronomy 32:4.
11. 1 John 4:10.
12. 1 John 1:5.
13. See Psalm 104.
14. C. S. Lewis, *A Grief Observed* (New York: Bantam Books, 1976), 76, 78.

CHAPTER 11: BUNGLING THE BAILOUT

1. B. E. Brown, "Coral Bleaching: Causes and Consequences," *Coral Reefs* 16 (June 1997): S129–S138, https://doi.org/10.1007/s003380050249.
2. A. W. Tozer, *The Knowledge of the Holy* (New York: HarperOne, 1978), 47.
3. Philippians 4:13.
4. Deuteronomy 31:8.
5. Matthew 28:17.

CHAPTER 12: THE BETTA HOSPITAL

1. Jean-Baptiste Raina et al., "Coral-Associated Bacteria and Their Role in the Biogeochemical Cycling of Sulfur," *Applied and Environmental Microbiology* 75, no. 11 (June 2009): 3492–3501, https://doi.org/10.1128/AEM.02567-08.
2. "The Beauty of Broken Objects," Smithsonian, August 7, 2020, https://www.si.edu/newsdesk/snapshot/beauty-broken-objects.
3. Gary M. Wessel et al., "Origin and Development of the Germ Line in Sea Stars," *Genesis* 52, no. 5 (May 2014): 367–377, https://doi.org/10.1002/dvg.22772.
4. Jeremiah 29:11.
5. Philippians 1:6.
6. Arthur Bennett, ed., "Earth and Heaven," in *The Valley of Vision: A Collection of Puritan Prayers and Devotions* (Edinburgh: Banner of Truth Trust, 1975), 203.

CHAPTER 13: SNUGGLING LIKE SEAHORSES

1. Jun Yu Chen, "Captive Breeding of Orchid Dottyback *Pseudochromis fridmani*: Reproduction, Ontogeny and Larval Feeding Regime" (thesis, James Cook University, 2021).
2. "Fish Pictures and Facts," *National Geographic*, accessed July 6, 2024, https://www.nationalgeographic.com/animals/fish. Originally published in National Geographic Society, *Animal Encyclopedia* (Washington, DC: National Geographic, 2012).
3. Jeremiah 29:11.
4. C. S. Lewis, *Present Concerns: Journalistic Essays*, ed. Walter Hooper (New York: HarperOne, 2017), 99.
5. 1 John 4:19.
6. Jeremiah 29:13.

CHAPTER 14: THE PROMISED GARDEN CITY

1. Emily Dickinson, "Hope Is the Thing with Feathers," Academy of American Poets, https://poets.org/poem/hope-thing-feathers-254.
2. "Genesis 2: Barnes' Notes," Bible Hub, https://biblehub.com/commentaries/barnes/genesis/2.htm; "Genesis 2: Cambridge Bible for Schools and Colleges," Bible Hub, https://biblehub.com/commentaries/cambridge/genesis/2.htm.
3. Francis, encyclical letter *Laudato Si'*, May 24, 2015, https://www.vatican.va/content/dam/francesco/pdf/encyclicals/documents/papa-francesco_20150524_enciclica-laudato-si_en.pdf.
4. Phylicia Masonheimer, *Every Woman a Theologian: Know What You Believe. Live It Confidently. Communicate It Graciously.* (Nashville: Thomas Nelson, 2023), 57.
5. Francis, *Laudato Si'*, May 24, 2015.
6. Garrett Hardin, "The Tragedy of the Commons," *Science* 162, no. 3859 (December 13, 1968): 1243–1248, https://www.jstor.org/stable/1724745.
7. Daniel J. Rankin, Katja Bargum, and Hanna Kokko, "The Tragedy of the Commons in Evolutionary Biology," *Trends in Ecology and Evolution* 22, no. 12 (December 2007): 643–651, https://www.sciencedirect.com/science/article/abs/pii/S0169534707002741.
8. Genesis 1:26; Psalm 89:14; 1 John 4:7.
9. Masonheimer, *Every Woman a Theologian*, 53.
10. 2 Corinthians 5:17.
11. Masonheimer, *Every Woman a Theologian*, 86.
12. Luke 24:39-43.
13. Luke 24:31, 36-37.

14. Timothy Keller, "The Song of Creation," from *Genesis—The Gospel According to God* sermon series, 12:57, October 15, 2000, Redeemer Presbyterian Church, Manhattan, New York, in *Gospel in Life*, podcast, October 8, 2015, 31:09, https://podcast.gospelinlife.com/e/the-song-of-creation.
15. Masonheimer, *Every Woman a Theologian*, 59.
16. John Mark Comer, *Garden City: Work, Rest, and the Art of Being Human* (Grand Rapids, MI: Zondervan, 2015), 248–249.
17. M. Eugene Boring and Fred B. Craddock, *The People's New Testament Commentary* (Louisville, KY: Westminster John Knox Press, 2009), 816.
18. Comer, *Garden City*, 250.
19. Genesis 3:22.
20. Francis, *Laudato Si'*, May 24, 2015.
21. "Chaotic Waters," episode 21 of the *How to Read the Bible* series in the *BibleProject* podcast, June 25, 2018, produced by Dan Gummel and Jon Collins, 1:11:34, https://bibleproject.com/podcast/series-h2r-p21-metaphor-e3-chaotic-waters.
22. Genesis 1:2.
23. Mark 4:35-41.
24. Revelation 5:13.
25. Revelation 21:22-23.
26. Linnie Marsh Wolfe, *Son of the Wilderness: The Life of John Muir* (Madison: University of Wisconsin Press, 2003), 144.

CHAPTER 15: BUBBLE-BRAINED

1. 1 Corinthians 1:25.
2. Charlotte Brontë, *Jane Eyre* (New York: Bantam Classics, 2003), 342–343.

CHAPTER 17: SILHOUETTES IN THE MAZE

1. Timothy Keller (1950–2023) (@timkellernyc), "The only person who dares wake up a king," Twitter (now X), February 23, 2015, 10:05 a.m., https://x.com/timkellernyc/status/569890726349307904.

CHAPTER 18: SOMEONE IS WATCHING

1. "How Should I Care for My Goldfish?," RSPCA Knowledgebase, updated May 1, 2024, https://kb.rspca.org.au/knowledge-base/how-should-i-care-for-my-goldfish.
2. Anna C. P. Locatelli et al., "Scientometric Analysis and Literature Synthesis of 60 Years of Science on the Atlantic Goliath Grouper (*Epinephelus itajara*)," *Journal of Fish Biology* 102, no. 4 (April 2023): 740–756, https://pubmed.ncbi.nlm.nih.gov/36635234.

CHAPTER 19: TENDING WILD GARDENS

1. Mark Turner, "*Cephalanthera austiniae*: Phantom Orchid," Turner Photographics, accessed July 22, 2024, https://www.pnwflowers.com/flower/cephalanthera-austiniae.
2. "Phantom Orchid," South Coast Conservation Program, accessed July 22, 2024, https://www.sccp.ca/species-habitat/phantom-orchid.
3. Lisa Boström-Einarsson et al., "Coral Restoration—A Systematic Review of Current Methods, Successes, Failures and Future Directions," *PLOS ONE* 15, no. 1 (January 30, 2020): e0226631, https://journals.plos.org/plosone/article?id=10.1371/journal.pone.0226631.
4. David E. Vaughan, ed., *Active Coral Restoration: Techniques for a Changing Planet* (Plantation, FL: J. Ross Publishing, 2021).
5. Kristen Hare, "He Devoted His Life to Understanding Florida's Coral Reefs," *Tampa Bay Times*, June 14, 2021, https://www.tampabay.com/news/florida/2021/06/14/he-devoted-his-life-to-understanding-floridas-coral-reefs.
6. Walter Jaap, email message to author, October 1, 2020.
7. A. W. Tozer, *The Pursuit of God* (Chicago: Letcetera Publishing, 2014), 95.
8. 1 Corinthians 15:58, NIV.

EPILOGUE: THE GREATEST HOPE

1. To read about some of these efforts, see Mote Marine Laboratory and Aquarium, https://mote.org/?s=Coral+Restoration; The Coral Nursery of Puerto Rico, https://www.coralnursery.org.
2. See Australian Institute of Marine Science, https://aims.gov.au/; Lauren Sommer and Ryan Kellman, "Scientists Are Breeding 'Super Corals.' Can They Withstand Climate Change?," NPR, April 1, 2024, https://www.kvpr.org/npr-news/2024-04-01/scientists-are-breeding-super-corals-can-they-withstand-climate-change.
3. The Coral Ark, https://www.thecoralarkinternational.org.
4. "Dry Tortugas National Park Acts to Preserve Genetic Diversity of Rare Pillar Coral," National Park Service, news release, August 26, 2001, https://www.nps.gov/drto/learn/news/dry-tortugas-national-park-acts-to-preserve-genetic-diversity-of-rare-pillar-coral.htm.
5. Acts 17:28.

about the author

Rachel G. Jordan is a professional marine biologist and lay theologian. A self-professed Jesus-loving coral nerd, she was raised in Idaho and has since traveled the globe chasing her passion for faith and science. She has an MS in marine biology and ecology from James Cook University (Australia), a BS in ecology from Seattle Pacific University (USA), and a certification in biblical studies from Bodenseehof Bible School (Germany). In addition to working as a coral biologist for the US National Park Service, Rachel has worked in marine aquaculture research, organic chemistry laboratories, veterinary research facilities, the pet industry, and a museum. When not diving or writing, she can be found reading C. S. Lewis, growing wildflowers, and exploring creation with her husband. You can also find her on Instagram @shorelinesoul or at rachelgjordan.com.

Tyndale | REFRESH

Think Well. Live Well. Be Well.

Experience the flourishing of your mind, body, and soul with Tyndale Refresh.

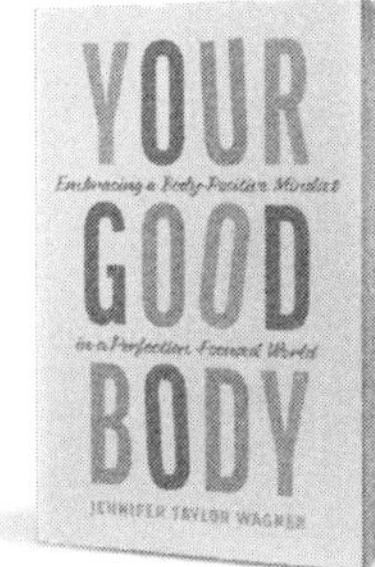

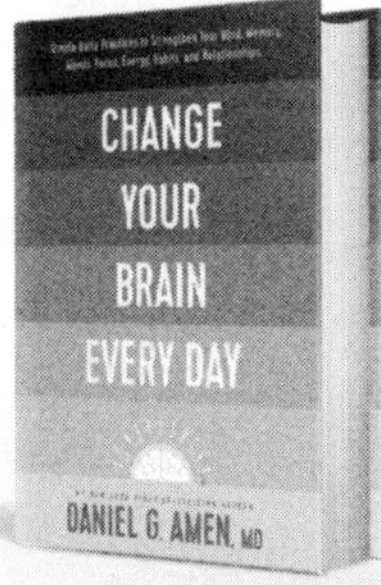

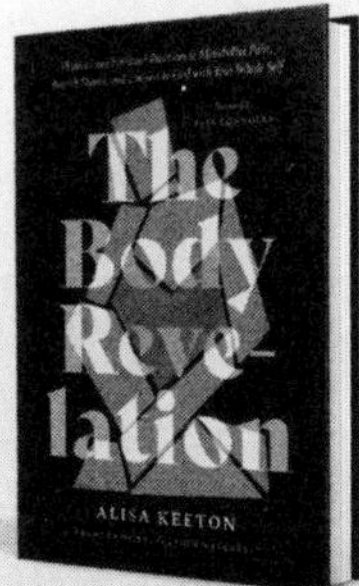

CP1841